From Biological Practice to Scientific Metaphysics

Minnesota Studies in the Philosophy of Science

ALSO IN THIS SERIES

Beyond the Meme: Development and Structure in Cultural Evolution
Alan C. Love and William C. Wimsatt, Editors
VOLUME 22

The Experimental Side of Modeling
Isabelle F. Peschard and Bas C. van Fraassen, Editors
VOLUME 21

*The Language of Nature: Reassessing the Mathematization of Natural Philosophy
in the Seventeenth Century*
Geoffrey Gorham, Benjamin Hill, Edward Slowik, and C. Kenneth Waters, Editors
VOLUME 20

Scientific Pluralism
Stephen H. Kellert, Helen E. Longino, and C. Kenneth Waters, Editors
VOLUME 19

Logical Empiricism in North America
Gary L. Hardcastle and Alan W. Richardson, Editors
VOLUME 18

FROM BIOLOGICAL PRACTICE
TO SCIENTIFIC METAPHYSICS

WILLIAM C. BAUSMAN, JANELLA K. BAXTER,
AND OLIVER M. LEAN, EDITORS

Minnesota Studies in the Philosophy of Science 23

 University of Minnesota Press

Minneapolis

London

Published by the University of Minnesota Press
111 Third Avenue South, Suite 290
Minneapolis, MN 55401-2520
http://www.upress.umn.edu

 Available as a Manifold edition at manifold.umn.edu.

ISBN 978-1-5179-1670-1 (hc)
ISBN 978-1-5179-1671-8 (pb)

A Cataloging-in-Publication record for this book is available from the Library of Congress.

Printed in the United States of America on acid-free paper

The University of Minnesota is an equal-opportunity educator and employer.

UMP BmB 2023

CONTENTS

TOWARD A SCIENTIFIC METAPHYSICS BASED ON BIOLOGICAL PRACTICE

WILLIAM C. BAUSMAN, JANELLA K. BAXTER, AND OLIVER M. LEAN

Despite the daunting complexity of biological systems and frequent failures, scientists have made significant advances in their ability to investigate, explain, predict, and manipulate those systems. A central task for philosophy of science is to understand what makes biological science successful in these endeavors. More and more, philosophers are realizing that this cannot be done by focusing solely on scientific theories. Instead, we must understand science as a rich system of *practices,* of which theorizing is only one kind. In turn, the fact that science is effective at learning about the biological world suggests that studying those practices might yield lessons about general features of the world that they successfully investigate (for examples, see Kaiser 2015; Kendig 2015; Meincke and Dupré 2020). In other words, successful scientific practice might inform *metaphysical* discussions.

The purpose of this volume is to explore and elaborate on these ideas—to investigate issues at the intersection between biology, practice-focused philosophy of science, and metaphysics. That is, it explores how a study of biological practice might contribute to a development of *scientific metaphysics*. As we discuss, definitions of metaphysics generally and of scientific metaphysics specifically are a matter of constant debate, and we do not aim to settle the matter here. In this volume, we want to distinguish *scientific metaphysics* from *metaphysics of science*. Metaphysics of science aims to characterize the entities, structures, and relationships that are at the core of or are assumed by a scientific theory or paradigm. Metaphysics of science is a modest project about the structure of the world as it is conceptualized and engaged with by scientists. By contrast, scientific metaphysics aims to make claims about entities, structures, and relationships of reality that go beyond a system of conceptualization and practice—claims about what the world is like, not just what scientists take the world to be like. The scientific metaphysics aimed for in this volume is an ambitious one. As we discuss later, the aim of this volume is to extend an analysis of scientific metaphysics beyond the scope of traditional accounts that primarily focus on theory and prioritize features like fundamentality, simplicity, and unity. It is a contention of the authors in this volume that any scientific metaphysics that pays insufficient attention to biological practice is at risk of misinterpreting the metaphysical significance of the science.

The views expressed in this volume about scientific metaphysics are diverse, but an overarching conclusion that emerges is that a metaphysics of the biological world informed by scientific practice is possible, potentially fruitful, and intrinsically interesting in its own right. For some contributing authors, this is a controversial goal. Because the claims of this volume depend importantly on the epistemic practices of scientists, there is a serious question as to whether and how we are ever justified in extending metaphysical claims beyond a paradigm. Our invocation of the Kuhnian concept of paradigm is meant to signal that scientific knowledge is often an expression of scientific traditions—defining features of scientific communities whose activities and thoughts are informed by shared conceptual frameworks, core questions, practical applications, techniques, instruments, and background beliefs (Kuhn 1962). Bausman's chapter represents the most explicit discussion of this problem, but many of the chapters touch on this issue. Other contributing authors are more comfortable with the project of scientific metaphysics. For example, chapters 6 and 9 provide metaphysical accounts that tentatively go beyond any particular paradigm.

This volume is the product of a three-year research grant sponsored by the John Templeton Foundation titled "From Biological Practice to Scientific Metaphysics" led by Principal Investigators Alan Love, C. Kenneth Waters, Marcel Weber, and William Wimsatt. The authors of this introduction—William Bausman, Janella Baxter, and Oliver Lean, also editors of the volume—were postdoctoral researchers on the project. The aim of this project was to investigate how an intensive study of scientific practices might inform, enrich, and correct metaphysical views. Many of the chapters in the volume defend arguments and ideas originally developed and presented at one of the numerous events of the project.

In this introduction, we situate the volume against related and contrasting subjects and approaches, discuss foundational issues surrounding its themes, and summarize the individual chapters.

1. BACKGROUND: BIOLOGICAL PRACTICE AND SCIENTIFIC METAPHYSICS

1.1 The "Turn" to Practice in Philosophy of Science

Philosophy of science in recent years has become increasingly interested in scientific practice. This shift in focus is sometimes called the *practice turn,* drawing a parallel with the linguistic turn in philosophy in the early twentieth century. To call an intellectual development a "turn" is a rhetorical device that endows it with the gravitas of revolution, with a well-defined and radically different "before" and "after." Whether or not this is strictly true of the turn to practice in science studies (Soler et al. 2014), it is useful as a label for a mode of inquiry historians and philosophers of science increasingly adopt in their analyses of science and scientific progress. Just as scientists often adopt different strategies, perspectives, and methods to advance their understanding of the world, philosophers and historians of science can "turn" to the study of practice to gain a more accurate and comprehensive view of how science works.

A recent concern among historians and philosophers of science is that scholarly communities have spent too much time in a theory-focused mode and not enough in a practice-focused mode (Chang 2014). What exactly does it mean to do philosophy of science with a focus on practice? The introduction to the volume by Soler, Zwart, Lynch, and Israel-Jost (2014) provides an excellent overview of this movement and its history. It is best to understand the practice turn as a reaction to several perceived problems with how philosophy of science (and science studies more generally, including the history and social studies of science) has tended to be done in the past. In the early analytical tradition, philosophical treatments of science focused on scientific theories from a narrow range of inquiries, particularly physics, thought to describe the "fundamental" features of the world. On this view, the relationship between theories, predictions, observations, and evidence is reconstructed using abstract sets of propositions related to each other by universal rules of logic and probability theory (Hempel 1945; Hempel and Oppenheim 1948). The job of the philosopher of science is to rationally reconstruct and critically evaluate this logic of science (Reichenbach 1938). Following World War II and the emergence of the Cold War, there emerged from this tradition a general, overarching attitude to what science is and what aspects of it are amenable to philosophical study based on the implicit assumption that scientists (qua scientists) are individual, ideally rational agents whose sole aim is attaining truth (Reisch 2005). Epistemology is not interested in the actual psychological thought processes or the external sociological contexts of people doing science (called the context of discovery by Reichenbach), but only in reconstructing the most logical way to arrive at a result and how to present it to other scientists (the context of justification) (Reichenbach 1938).

It is this inherited attitude to science and its relation to philosophy to which the practice turn is a reaction. While many valuable insights came from this traditional approach to philosophy of science, it also led to many important facets of science being overlooked. For example, the approach's emphasis on universal laws of nature makes it hard to accommodate almost all areas of inquiry beyond physical mechanics (Mitchell 2003; Wimsatt 2007; Currie 2018). Moreover, as many philosophers, sociologists, and historians of science have argued, this focus leaves unaddressed many aspects of scientific inquiry that are crucial to scientific progress (Pickering 1992; Nersessian 2012; Leonelli 2016). Theoretical and conceptual success is intricately intertwined with epistemic activities including intervening, classifying, data gathering and organizing, modeling, and doing statistics. Scientific success is often achieved through an iterative process whereby new conceptual and technical developments help build upon and correct previous understandings and methods of interacting with the world (Chang 2004).

In short, practice-focused philosophers typically view science as a richly interconnected system of activities, skills, strategies, and background knowledge by means of which scientists investigate their domains for a variety of purposes. In particular, they avoid viewing science purely in terms of its theories. That isn't to say that theories are not important to science—they certainly are, though not always as much as supposed. Rather, it is dubious whether an adequate understanding of scientific theories can be had when studied in isolation from the investigative context in which they arise. This context shapes the structure a theory eventually takes. And when theories are important, they are important insofar as they affect how science is *done*— what questions are asked and pursued, what experiments are performed, how data are interpreted and evaluated, and so on. This contrasts with a theory-focused approach that grants those practical aspects a peripheral role, in which practice is just the means by which theories are developed and evaluated. A practice-focused approach takes the opposite view: theories get their significance from the wider system of practices in which they are embedded. Theories are just one of the tools in the scientists' toolbox, and just like the others, what matters is how they are used.

Along with contextualization of theory and focus on activity, practice-focused philosophers are typically interested in the complex realities of science rather than the abstractions and idealizations that were typical of traditional approaches to studying science. Most famously, Kuhn (1962) criticized the traditional view of scientific theory change on the grounds that it didn't fit with how scientific revolutions actually happen in history. Mistakes like these have been made, it is said, because of philosophy's failure to pay proper attention to how science actually works. Disabusing us of the idea that science is centered on testing and refuting hypotheses, Kuhn's portrayal of "normal science" drew attention to another aspect of theoretical work; namely, the articulation and extension of the theory's central hypotheses and exemplars. Since then, focus has also been drawn to the role of experimental work (Hacking 1983). A more drastic shift has been away from the traditional view of scientists as isolated, ideal agents and toward viewing them as encultured communities of imperfect human organisms with rich inner and outer lives and working with finite material and cognitive resources. Philosophers of scientific practice are often interested in how science succeeds within these constraints rather than imagining how it would or should be without them (Wimsatt 2007; 2023). In particular, philosophers of scientific practice often emphasize the idea that science is and has always been a purposeful activity, performed against a background of goals and values. This aspect overlaps strongly with feminist philosophy of science, which critically investigates those goals and values and the scientific ideas they support (see later).

Because of this interest in science as a complex system of activities, practice-focused philosophers often incorporate scientific methods into their research to understand that complex system. Practice-focused philosophy therefore overlaps strongly with *empirical philosophy of science,* which incorporates tools such as ethnographic studies of scientific communities (Nersessian 2012; Leonelli 2016; Kaiser and Trappes 2023), psychological experiment (Griffiths et al. 2009; Lombrozo 2010; Riesch 2015; Machery et al. 2017), or digital analysis of scientific literature (Lean et al. 2021; Overton 2013; Pence and Ramsey 2018; Mizrahi 2020). The hope with these methods is that philosophical preconceptions about science can be refined or overturned with observation, thereby enriching the picture of science on which our philosophy is based.

In summary, we take philosophy of scientific practice broadly to study science *in practice* instead of science *in theory*—rather than as an idealized phenomenon abstracted from its complex context. We will see various examples of how attention to practice shifts and expands philosophical discussions about science in the rest of this introduction and throughout the volume.

1.2 Scientific Metaphysics

The overarching aim of this collection is to further the cause of enriching the philosophy of science through the study of scientific practice by addressing a particular and largely overlooked aspect. Specifically, it explores how taking a practice-focused view of science can contribute to questions in metaphysics.

Metaphysics as a subject is notoriously hard to define (Sullivan and Van Inwagen 2020), but it is typically said to concern the nature of reality in some general, ultimate, or fundamental sense. Examples of subjects usually classed as metaphysical include existence, space, time, causality, identity, possibility and necessity, parthood, free will and determinism, natural kinds, and the place of mind or consciousness in nature. These subjects might be seen to concern what some consider the general categories of being and how they relate to each other. It is a common view among metaphysicians that the discipline is prior to or independent of the results of the sciences (Bealer 1982; Lowe 2002; Hawley 2006). Of course, most of the metaphysical subjects just listed are relevant to and used in the sciences. However, some, such as L. A. Paul (2012), argue that the sciences merely study the *instantiations* of these general categories, whereas metaphysics studies those categories *as such*. Similarly, Lowe (2002) argues that metaphysics is prior to science because it provides the very conceptual framework within which scientific ideas are expressed (see Kincaid 2013 for extended discussion).

The idea that metaphysics is independent from and even prior to the sciences has undergone waves of criticism, from Hume to the logical empiricists of the twentieth century (Carnap 1950; Creath 2023). This kind of criticism has once again gained new life in philosophy in recent years. One source of skepticism lies in the a priori methods those approaches to metaphysics are said to rely on, such as appeal to intuition (Ladyman and

Ross 2007; Kincaid 2013; Bryant 2017) (though see Bennett 2016). Critics argue that there is no reason to expect reality to conform to our intuitions, which are highly variable and contingent and which have evolved to serve aims unrelated to objective truth. Claims of this kind are supported by empirical work in experimental philosophy, which shows cross-cultural variations in intuitions (e.g., Rose 2020), thereby challenging the universality and authority of the intuitions on which canonical philosophy is based. In contrast, Paul (2012) defends metaphysics by arguing that it shares with science a methodology based on inference to the best explanation (IBE)—that is, a preference for theories with virtues such as simplicity and generality. Since science uses IBE and is successful in doing so, this indirectly vindicates metaphysics, which uses the same methods. However, aside from questions about whether the legitimacy of IBE survives the transition from science to metaphysics, it may in fact be false that IBE is as generally abundant in science and especially biology as Paul takes it to be (Novick 2017). If so, Paul's indirect justification for metaphysics is unfounded. Alternatively, one might regard some arguments for scientific metaphysics (in this volume and elsewhere) as implicitly resting on a kind of IBE logic, interpreted as arguing that metaphysical conclusions are what best account for the success of some particular set of scientific practices.

The preceding criticisms target the *justifications* that metaphysicians provide for their claims, arguing in various ways that those methods of justification are inadequate. Other criticisms run deeper: for some, claims about things that lie outside empirical experience, even in principle, are not just unsupported but meaningless. This kind of argument has roots in Hume's famous distinction between matters of fact and relations of ideas and was famously elaborated formally by the logical positivists in their principle of verification. It is also a major theme in American pragmatism—an important intellectual connection between Hume and the logical positivists. C. S. Peirce, for example, claimed that the Catholic doctrine of transubstantiation was not just false but meaningless precisely because it posits a fact that lies beyond all possible experience: "To talk of something as having all the sensible characters of wine, yet being in reality blood, is senseless jargon" (Peirce 1878/1935). Peirce, like many whose work he inspired, believed that metaphysics properly understood did not lie outside experience but worked *for* experience. In other words, metaphysical ideas serve to structure and organize our interactions with the world in useful ways rather than being about things that make no difference, even in principle, to those interactions.

(See Lean 2021 for a discussion of this idea in relation to contemporary biological data practices.)

In summary, scientific metaphysicians hold that if metaphysics is to be a meaningful and worthwhile pursuit, it should be based as far as possible on lessons from the sciences rather than just on a priori reflection. This view is also driven by a positive argument: since science investigates reality and is highly successful at doing so, its success-making features should be relevant evidence for what that reality is like. However, acknowledging the relevance of science to metaphysics leaves open exactly what the relationship between the two should be; that is, exactly what aspects of science are relevant to metaphysics and in what way (Hawley 2006; Kincaid 2013; Chakravartty 2013; Bausman 2023; Creath 2024). What is needed, in short, is an account of how science relates to the world such that the former can serve as evidence for the latter and that offers recommendations for how to gather and use this evidence in metaphysical discourse.

2. A METAPHYSICS BASED ON BIOLOGICAL PRACTICE?

This volume explores the intersection between the two philosophical movements outlined previously—that is, between scientific metaphysics and a practice-oriented philosophy of science. As mentioned, scientific metaphysics is motivated by science's success and considers this to be worth taking seriously when we ask what the world is like in metaphysical terms. For the most part, however, this project has restricted its attention to successful scientific *theories*. But from the practice perspective, theory is only one aspect of what makes science successful. This suggests that the practice perspective may bring attention to a vast collection of success-making factors in science that metaphysics can potentially draw on. For example, Baxter's chapter represents how a practice-based approach to scientific metaphysics opens up new questions about the metaphysical commitments of things like genome databases—a practice that has only recently received philosophical attention.

What's more, the theories on which scientific metaphysicians have focused have largely been those of physics. Ladyman and Ross (2007), for example, argue that unifying the sciences by relating their theories to those of fundamental physics is what a properly naturalized metaphysics should be for. If that is what one means by metaphysics, then either failure to relate a science to fundamental physics or finding a disunity between the sciences would invalidate their entire metaphysical project. However, there are other ways to understand the project of metaphysics that still recognize the importance of paying attention to science. On some of those ways, the biological sciences have much to offer in metaphysical discourse in their own right, in ways that don't necessarily involve relating those sciences to fundamental physics.

Biology often fails to fit metaphysical theories that are developed in relation to physics, and so it offers an important set of case studies for testing the general validity of those theories. One example is the relationship between classical and molecular genetics, which fails to fit standard models of reduction (Waters 2008; Weber 2006). Second, scientifically friendly metaphysicians might be interested in exploring what (if any) metaphysical conclusions can be drawn from traditional biological questions—questions about natural selection, intentionality and the mental, biological individuality, species, and biological information. These concepts apparently concern aspects of our reality and can be interpreted as metaphysical. They have also often been seen as metaphysical historically. What's more, those aspects are of particular importance to human society. Hence they may be worthy of close attention by metaphysicians who sympathize with John Dupré's sentiment that "biology is surely the science that addresses much of what is of greatest concern to us biological beings, and if it cannot serve as a paradigm for science, then science is a far less interesting undertaking than is generally supposed" (1995, 1). Overall, the kinds of phenomena that the biological sciences study are closer to home than and also qualitatively different from those of physics. For philosophers interested in exploring metaphysical views of the everyday biological reality we encounter, this volume is a step in that direction.

Why has this potential gone largely unrealized? Because, in general, practice-focused philosophers have taken a negative or debunking attitude toward metaphysical interpretations of scientific claims rather than aiming to refine and improve them. For example, against those casting questions about typology as metaphysical and connected to essentialism, Love (2009) argues that typological thinking in developmental biology is simply an epistemic strategy that can yield different categorizations according to different goals. Metaphysical interpretations of science, such as determinism and essentialism in biology, have also long been the target of important criticisms from a feminist perspective (Longino 1990; Dupré 1995; Gannett 1999; Haslanger 2016). What this means is that in addition to scientific metaphysics largely overlooking practice, philosophers of scientific practice have largely overlooked or been broadly critical of metaphysics. These two realms of inquiry have largely been mutually exclusive or even hostile.

As valuable as these critical views of metaphysics in science are, one may also ask whether it is possible to develop a *positive* metaphysical project that adequately accommodates their criticisms. It is time to drop the implicit assumption that showing the aims and values of scientists who generated a result is enough to undermine any further metaphysical interpretation of the result. There are, to be sure, already examples of philosophical work that do appeal to biological practice in advancing metaphysical claims. Probably the clearest example of this is John Dupré's *The Disorder of Things* (1995). There, Dupré identifies and criticizes a number of metaphysical precepts underlying discourse about science; namely, essentialism, reduction, and determinism. All of these assumptions are misplaced, he argues, because they do not square with observations about how scientists (biologists in particular) actually go about investigating the world. If we reject these precepts—as he argues we should—we instead find a metaphysics based on disunity, variety, and change. More recently, Dupré and others have defended a metaphysics based on processes over substances (Dupré and Nicholson 2018). As another example, Joseph Rouse (2002) argues that scientific practices hold the key to dissolving the conflict between the causes of nature and the norms of the social world. His arguments center around understanding humans as social organisms interacting purposefully with their environment and scientific practice as, in effect, a special case of niche construction.

Most of these works appeal to scientific practice in order to advance a particular metaphysical thesis or provide different practice-based angles on a particular metaphysical issue. The present volume's broader aim is to showcase ways in which biological practice might inform a variety of metaphysical discussions as well as explore foundational questions about whether and how inferences from biological practices to metaphysical conclusions are possible.

Ultimately, demonstrating the possibility and the value of a metaphysics based on biological practice means doing the work of addressing individual metaphysical questions from that perspective. In the next section, we discuss particular questions and concepts that are typically characterized as metaphysical and that are addressed in more or less explicit ways throughout the chapters of this volume.

3. THE NATURE OF METAPHYSICS AND SCIENTIFIC REALISM

As we've seen, there is no agreement on what counts as metaphysics: For some, metaphysics is by definition distinct from and prior to science or to empirical inquiry generally. For others, metaphysics is a post-scientific exercise of unifying scientific theories by relating them all to a common theoretical foundation such as fundamental physics in order to discover what there is. The works in this volume depart in important ways from most approaches to metaphysics: For one thing, they emphasize practice over theory. For another, they generally treat biological sciences as metaphysically interesting in their own right, independently of questions of whether and how they relate to fundamental physics. Of course, one may simply deny that these are works of metaphysics on the basis that they fail to meet one's preferred definition of the term. We aim to provide reasons why we take the approaches in this volume to be worthy of the label, even if some will not agree. Despite many differences in approach in the chapters of this volume, we see among them a common thread underlying their respective subject about what a metaphysics from biological practice involves. Here we articulate this idea and place it in context with other views.

In our view, justifying a philosophical project as metaphysics involves at least two things. First, it should be recognizably *meta*-physical rather than "merely" scientific. Of course, this doesn't mean that the two are strictly different pursuits; it may be that doing science sometimes means doing metaphysics in the process, or vice versa. (Lean [2021] takes a view to this effect in regard to digital ontologies in biology, for example.) While scientific metaphysics is by definition informed by science, it should contribute something to our understanding that is not contained in the explanatory or investigative reach of the science. Importantly, it must have an answer to the question of what a metaphysics based on science can add to our understanding that the science itself cannot provide without presuming to go "around" or "beyond" the science, which naturalistic philosophers consider to be problematic. Sometimes metaphysicians ask different questions or investigate different domains than scientists, other times it is their approach to investigating the same questions as scientists, and other times the same work can be considered scientific and metaphysical. For example, Ladyman and Ross (2007) satisfy this criterion by viewing the role of metaphysics as *unifying* diverse scientific theories. Even if such a project were to qualify as a science in itself, this second-order, unificatory goal also makes it meta-scientific while still adhering in important ways to scientific results. Other metaphysical work inquires into the implications of a part of science for the nature or reality of something. This inquiry must not be seen as extra-scientific tout court, only extra- this specific part of science, as when psychologists ask about the implications of evolutionary biology on modern human psychology.

Second, as well as being distinct from the sciences, a metaphysical project should be distinct from epistemology: there should be a sense in which the questions one is asking about science are not just about the scientists' concepts, beliefs, practices, or whatever but also about the nature of the domain they investigate. It is on this point that practice-focused philosophers have often ignored or actively distanced themselves from metaphysics, since they often take their subject matter simply to be scientific practices and social epistemology and the like rather than an attempt to articulate the nature of the world that those sciences describe. Hence the challenge for practice-focused scientific metaphysics is to justify the metaphysical import of one's claims without ignoring or glossing over the epistemic aspects, since (on the scientific metaphysics view) it is precisely those epistemic aspects and their success-making features that underwrite science's relevance to metaphysical questions.

We take most of the work in this volume to meet these criteria in the following way. First, the metaphysics undertaken here differs from scientific inquiry itself because it studies those practices in which scientists are engaged. It therefore shares its direct object of inquiry with the rest of science studies—that is, with history and social studies of science (Soler et al. 2014). An important principle of science studies is that many important aspects of science are implicit or unconscious and not well understood or even necessarily known to the scientists themselves. It is therefore a task of science studies such as philosophy to make those things explicit so that they can be understood and critically analyzed. One example is the concept of repertoires, which was developed by philosophers and sociologists to understand how successful systems of scientific practice are reproduced in new contexts (Leonelli and Ankeny 2015; Ankeny and Leonelli 2016). Note that this is a different way to work "above" science than analytic approaches to metaphysics: rather than attempting to go around science or to establish its a priori foundations, as some metaphysicians have aimed to do (see Creath 2023; Ereshefsky and Reydon 2023), the chapters in this volume stand apart from science in that they take science-in-practice as their direct *object of inquiry*.

This itself is insufficient to establish such a project as metaphysical, rather than simply metascientific or philosophical in general, because of the second criterion. That is, we also need a justification for why this sort of project concerns not just the scientific practices themselves but also the world outside those practices. The reason is that we are centrally concerned with what makes a science *successful,* and that success is partly shaped by things that are extrinsic to science itself; in particular, the world it investigates. Note that this is simply the basic motivating principle of scientific metaphysics in general—that science is relevant to metaphysics because it is successful—generalized to make both theoretical and nontheoretical practices metaphysically relevant. Many of the approaches in this volume entertain the possibility that scientific practices are successful because they *fit* aspects of reality that they encounter. Here, "fit" can be understood in engineering terms: those practices exhibit features that make them well designed to achieve particular aims in the particular environment they work in.

It is worth noting that the idea of strategies fitting their environments has become a commonplace principle in the science of machine learning—arguably a form of applied epistemology. There, it is widely accepted that there is no single best method for collecting and interpreting data that is optimal regardless of the world's structure (Wolpert and Macready 1995; Korb 2004); in other words, the success of epistemic strategies for learning about the world depends on what the world is like. The inseparability of metaphysics and epistemology is far from a new idea, then, though we take it to be particularly important here: while our focus is on successful epistemic practices, it is critical to remember that those practices are successful when and because they work well in their *external environment,* as Herbert Simon (1996) put it. We cannot ignore the role of the world in enabling successful science any more than we can ignore the role of the environment in biological adaptations.

In short, we take the work in this volume to adopt, implicitly at least, what can be called a *functional* approach (Woodward 2014; Woody 2015). That is, it supposes that scientific practices have goals and purposes, sets out to determine what those purposes are, and therefore critically evaluates the extent to which they are successful in their goals. Importantly, since an overarching goal of science is to investigate the world, what makes it successful has at least something to do with what the world is like. This view is most explicit in Ereshefsky and Reydon (2023), in which classification systems are argued to pick out "natural" kinds when the success of those systems is "grounded" in features of the world. Similarly, Lauren Ross (2023) effectively treats epistemic strategies in psychiatric genetics as solutions to the complexity of their domain of inquiry. Bausman (2023) aims directly at an analysis of this kind of inference by comparing it to similar arguments in biology that are based on claims about adaptation.

These questions about the relationship between science, epistemology, and metaphysics closely connect to the issue of scientific realism. In the broadest terms, scientific realism amounts to the idea that science aims at describing or getting at reality. Traditionally, scientific realism has been formulated in terms of true theories—or theories that describe a set of justified true beliefs (Van Fraassen 1980). Truth is standardly understood as a correspondence between what our propositions say and how the world is. For scientific realists, the success of those theories is best explained by the idea that they are at least approximately true. However, it is difficult to evaluate systems of scientific inquiry on the basis of truth from a practice-based approach to philosophy of science because much of scientific knowledge is non-propositional and, thus, is neither true nor false (Baird 2004; Waters 2014; 2017; Chang 2017). Instead, many scientific practices are better understood as *knowledge-how or knowledge-as-ability* rather than *knowledge-that or knowledge-as-information* (Chang 2017); they involve skills and abilities to act. This means that if we are to engage in reasoning about reality by observing successful practices, it should be based on a different or more general relation than truth. One way to do this may be to adapt the idea of truth-makers—the facts or states of affairs that impart truth on statements about the world—and instead talk about *use-makers,* or states of affairs that impart usefulness on scientific practices. As Otto Neurath once wrote, "statements are compared with statements, not with 'experiences,' not with a 'world' nor with anything else" (1931/1983, 53). We say that scientific statements are truth-apt and the world supplies the truth-makers. Following this, we should say that scientific practices are *use-apt* and the world supplies the *use-makers.* When a practice fits an aspect of the world to a high degree, those aspects of the world make those practices useful. Some philosophers might wish to defend a version of scientific realism constructed around a more sophisticated notion of truth, but this needn't be our only way of thinking about scientific realism.

In contrast to a focus on truth, we could instead explore how to think about scientific realism in terms of the broader idea of progress. Science makes progress whenever it helps us learn something new—whether it be a piece of propositional or practical (non-propositional) knowledge—from reality. Reality is not subject to our whim. We do not make reality by simply thinking or imagining. Instead, reality resists or frustrates our efforts to carry out our aims. Practical know-how is whatever skills and beliefs are needed to successfully engage in epistemic activities in the world. We are sympathetic to what Hasok Chang calls active scientific realism. Active scientific realism is a commitment to continually seek greater contact with reality for as many different aims as we may wish (Chang 2012). Importantly, our proposal has some important differences with Chang's view. Chang is quite skeptical of what we are calling scientific realism. Instead, he maintains that metaphysics of science is the best philosophers can achieve. At least some of the authors in this volume are more hopeful that a scientific metaphysics grounded in progress is possible.

Setting aside the status of truth and scientific realism, we can distinguish between different types of projects a philosopher of science interested in metaphysics might pursue. Following Roe (2015), Waters (2017), and others, one type of project a philosopher of science might engage in is what we call the *metaphysics of science*. This project examines the metaphysical assumptions and commitments made by a system of scientific inquiry and differs from how others have defined the project (Mumford and Tugby 2013). For example, for the developers of the kinetic theory of gases, heat is the energy of molecules in motion. Kuhn's (1962) analysis of paradigms or disciplinary matrices includes metaphysical commitments of this kind. Relative to the system of inquiry within which scientists work, the relationships and entities that characterize the inquiry can be thought of as an "internal ontology" (Gillet 2021). The metaphysics of science is a descriptive and interpretive project toward understanding how a system of scientific inquiry operates internally—according to its own logic and assumptions. As such, it need not be thought to say anything about what the world is like outside a system of inquiry without further argument. This is in contrast with what we will call *scientific metaphysics*. This project attempts to get outside a particular system of scientific inquiry to say something about what the world is like by drawing on the practice and results of science.

The metaphysics of science versus scientific metaphysics distinction is independent of an interpretation of the claims of scientific metaphysics. Scientific metaphysics claims will still be open to interpretation and different schools will interpret them differently. For example, "Heat is not a substance but molecular motion" might describe states of affairs, or it might express an attitude toward life (Carnap 1931). It might be seen as truth-apt or use-apt. Either way, metaphysical claims of this sort require substantial justification beyond internal coherence and success (Chakravartty 2010; Morrison 2011; Chang 2014; Bausman 2023). For the success of a system of scientific inquiry can be due to a variety of nonscientific factors—such as the interests, values, and social structure of a scientific community.

Importantly, on our conception, metaphysics of science is a necessary first step toward a scientific metaphysics. Before we can know whether we are justified in concluding that atoms, genes, gravitational fields, or economies have an ontological status outside the systems of inquiry from which they are described, we have to know how to interpret and understand these objects from within their systems of inquiry. Arriving at the metaphysics of a scientific project can require significant philosophical reflection on the vocabulary, activities, and social structures informing an area of inquiry. Several chapters in the volume are engaged in metaphysics of science projects. For example, Baxter's contribution aims to clarify and defend the GenBank gene concept as a useful structure for the purposes of organizing and disseminating genomic information through the National Center for Biotechnology Information's digital database system. Kaiser and Trappes clarify the NC^3 mechanisms at work in a collaborative project in which they participate titled "A Novel Synthesis of Individualization across Behavior: Niche Choice, Niche Conformation, Niche Construction." These authors maintain that the GenBank gene concept and NC^3 mechanisms have ontological character internal to the systems of inquiry that investigate and develop them. However, as Kaiser and Trappes note, we should understand the metaphysical status of such entities as "provisional" for the analyses offered in these chapters do not supply the justification required for a scientific metaphysics.

Can philosophers of science defend a scientific metaphysics, and if so, how? They have several options available to them. One possibility is perspectival realism (Giere 2010; Chirimuuta 2016; Massimi 2018). On this view, a propositional claim P is true so long as it is true according to the truth conditions of its scientific system of inquiry. For example, from the perspective of modern chemistry the statement, "water is H_2O" is true and the statement "water is HO" is false; however, from the perspective of Dalton's system, the latter statement is true (Chang 2014). An objection to perspectival realism is that it provides no means for evaluating which systems of inquiry "get things right" in a more absolute sense that antirealist accounts of science fail to supply. One proposal for resolving this problem is to evaluate the adequacy of different systems of inquiry not on the truth of propositions but on various dimensions of success—such as explanatory and predictive power, testability, manipulability, and so on (Massimi 2018). Success can be measured along multiple parameters and, unlike truth, is a matter of degree. Thus, we can compare different systems of inquiry. For some philosophers of science, greater degrees of success might be taken as an approximation for truth, whereas others, like Hasok Chang, can reject this view and just treat success as the primary aim of science.

Another possibility is to take Bill Wimsatt's (1994) view of robustness as a criterion for the reality of objects and properties. Robustness here means that an entity is accessible in a variety of independent (in the sense of probability of failure) ways. Therefore, if an entity is detectable or manipulable using independent instruments and predictable from independent theoretical assumptions, it is (probably) real. Comparing Wimsatt's view to perspectival realism, Wimsatt takes the "objectivist" step using robustness: if there are many independent reasons (models, theories, experiments, instruments, etc.) for thinking that heat is not a substance but molecular motion, then heat is molecular motion.

Finally, in keeping with the broadly pragmatic approach of this volume, we should consider not just what metaphysics *is,* in our view, but also what it is *for.* As we've seen, one interpretation of the role of scientific metaphysics is to unify the sciences with fundamental physics in order to place constraints on acceptable theories in the special sciences. While we do not feel the need to criticize this view, we believe there are other valid motivations to do scientific metaphysics. As discussed, many metaphysical questions raised about the natural world are of considerable significance to humans' understanding of ourselves and of our relationship to our reality more widely. Virtually every facet of human society—ethics, law, economics, religion, the arts, politics, and so on—involves commitments about the nature of the world we inhabit. How much and in what ways are those commitments accountable to the results of the sciences? The question of the relationship between science and metaphysics is, in fact, a question about the authority of science and its place among other types of inquiry and in society as a whole.

We see an important role for scientific metaphysics in offering clarity about the nature and extent of science's authority in our broader conceptions about the world. Ladyman and Ross vehemently deny the idea that the role of philosophy is "to try to make the world as described by science safe for someone's current political and moral preferences" (2007, 6). We wholeheartedly agree with this sentiment—science can and should be invoked to challenge moral and political beliefs if those beliefs are based on ignorance, misunderstandings or overinterpretations of scientific output. However, we see a great deal of work left to be done to determine exactly when and in what ways the results of science can be brought to bear on those other facets of human life. For example, to what extent should society's view of sex be based on biology? To answer this question, one must consider the role that sex categorizations play in biological practice; for example, in comparing and explaining reproductive strategies. From that, we can consider the various purposes for assigning people to sex categories in human society and then ask to what extent those purposes overlap or interact with those of biological practice. One can reasonably accept the importance of biological sex to biologists and the "reality" of biological sex—in other words, that sexes are natural kinds in Ereshefsky and Reydon's sense—while questioning whether society should use sex categories in the same way that biologists do. Looked at in this way, the insistence in the public sphere that "sex is biological" is in effect the claim that all of society's institutions should adopt biology's standards of sex classification. It is therefore revealed to be a normative social claim rather than a factual scientific one and should be justified and criticized accordingly.

The real world value of a practice-oriented scientific metaphysics extends even beyond challenging moral and political beliefs founded on misinterpretations of scientific knowledge. By carefully describing the conceptual and technical developments at the heart of scientific practices, philosophers position themselves to embark on a kind of project Hasok Chang calls complementary science (Chang 2004). Complementary science involves advancing alternative conceptual schemes that have been neglected, forgotten, or suppressed by scientific institutions but that have the potential to advance human understanding of the world in novel ways. Complementary science is driven by normative considerations. At this stage of inquiry, the philosopher of science may embark on what Sally Haslanger (2000) calls an analytical project, whereby one asks what work we want our concepts (such as race and gender) to do for us. Descriptive works in the history and philosophy of science necessarily build the foundation for and thus must come prior to complementary science. Philosophy of science that carefully details the conceptual and technological developments of scientific inquiry helps uncover the contingent choices scientific actors have made in the past. This is no small lesson to glean from the study of scientific metaphysics. In clarifying the ways scientific systems of practice are partially informed by the choices scientific communities make, philosophers of scientific metaphysics help identify ways in which scientific systems of practice could be otherwise. In this way, philosophers of science can bring epistemological, social, political, and ethical considerations to bear on the ways scientists engage with the world. Yet, before philosophers of science can take up complementary science, extensive descriptive work must be done. That is what much of the work in this volume strives to do.

This volume represents the view that attention to scientific practices is critical to questions about how science should constrain our wider belief systems, since it is all too easy to misinterpret the significance of theoretical claims in science if we fail to consider the practical context in which those theories operate. In short, one can only properly interpret what science *says* about the world by first understanding what science *does*.

4. CAUSALITY

There is a deep tension in the notion of causality. On the one hand, it appears to be essential to our understanding of the world: the only way we can reason about the unobserved is to posit connections behind our observations that will be repeated in future. On the other hand, we never directly observe causal relationships, only regular co-occurrences. In other words, no causal claim is logically entailed by any possible observational data. This problem is especially pressing from the point of view of scientific metaphysics. Bertrand Russell (1913) famously argued that there is no causality in fundamental physics, since its laws are all symmetrical whereas causality is asymmetric (see Ladyman and Ross 2007 for critical discussion). Importantly, the special sciences such as biology appear to be deeply committed to causality, as evidenced by their prominent appeal to causal explanation, process, mechanism, and other causal concepts. If we insist on getting our metaphysics from science and yet science merely *presumes* causality where it uses it at all, what does this mean for the idea of causality in a scientific worldview?

The shift of perspective to scientific practice has revitalized philosophical discussion of biological causation in recent decades. From the practice perspective, rather than beginning with questions about what causality "is" and then looking to scientific theories to tell us the answer, we can instead begin by analyzing causal *reasoning* and the purposes it serves in the various practical goals of science. This view of the relationship between philosophy and practice in relation to causality is neatly summarized by Dewey: "The first thinker who proclaimed that every event is effect of something and cause of something else, that every particular existence is both conditioned and condition, merely put into words the procedure of the workman, converting a mode of practice into a formula" (1958, 84). The role of philosophy, from this perspective, is to make explicit and to codify those practices that constitute science's understanding of causation.

This practice-first attitude is what motivates *interventionist* theories of causation such as that of Woodward (2003), which several chapters in this volume draw on. According to interventionism, causal reasoning is closely tied to purposes of manipulation or control: to reason about causes is to reason about how to change outcomes through actions. In other words, causal reasoning reveals the "levers" or "handles" that can be exploited (in principle) to change things in our environment. Interventionism in philosophy connects closely to sophisticated scientific work in statistics and artificial intelligence, which has developed an interventionist logic of causal reasoning (Spirtes et al. 2000; Pearl 2009). Interventionism holds that *scientific* progress has partly consisted of developing more sophisticated and elaborate ways of investigating and reasoning about these causal levers; this is what it is to develop better explanations (Ismael 2013; 2017; Woodward 2014; Ross forthcoming).

Interventionist approaches to causation in the sciences are especially useful when applied to biological and biomedical sciences. Scientific principles about the workings of biological systems are exception-ridden, contextual, and contingent, which makes it difficult to view causal reasoning in biology in relation to laws. Causal reasoning in biology therefore offers a rich testing ground for ideas about high-level and top-down causation (Wimsatt 1976; Craver and Bechtel 2007; Woodward 2008; Green and Batterman 2017; Weber 2006). Woodward argues that as well as distinguishing causes from non-causes, biologists also distinguish one cause from another along various dimensions (Woodward 2010). For example, one cause-effect relationship is more *stable* than another if it holds under a wider variety of background conditions, or under a wider range of manipulations. Importantly, these distinctions *matter to us* because they afford different types and degrees of control and prediction. This line of thinking has recently been applied to the controversy about the privileging of genes among other causes in development (Weber 2006; Waters 2007; Griffiths and Stotz 2013; Ross 2020; Baxter 2023).

Another key part of understanding the role of causal reasoning in biology is to understand how it connects to other kinds of reasoning. For example, one closely related concept is that of *mechanism*: biologists often explain biological phenomena by citing the entities and/or activities that interact to produce those phenomena (Machamer et al. 2000; Craver 2007; Glennan 2017) (though see Ross 2018). There has been much discussion in philosophy about what mechanistic explanation amounts to, how much of biology is concerned with it, and how the notion relates to other concepts such as that of interventionist causes discussed previously. Lean's (2023) present contribution discusses the relationship between causality and the concept of *information* that is widely used throughout the biological sciences. In keeping with a functional approach to understanding scientific practice, understanding the role of one type of practice means understanding how it connects to others.

For some, especially those engaged in questions of causality in analytic philosophy, these practice-focused treatments of causation may not count as metaphysical at all. It may be argued that these "merely pragmatic" considerations are quite irrelevant to the concerns of metaphysicians, who are instead interested in causality's metaphysical basis or grounding (Woodward 2015; 2017). This is of course true if one's definition of metaphysics makes it essentially irrelevant to practical concerns. However, as we have stated, a major aim of this volume is to explore the possibility of a metaphysics that *is* informed by practice. In fact, when we recognize scientific practice as the engine of success, the very possibility of a scientific metaphysics (based on our most successful science) implies the relevance of practical matters to metaphysical problems. In the case of causation, it is not unreasonable to suppose that the way a science uses causal reasoning might have something to do with what the world is like. In the case of biology, we might coherently ask whether there is something *about* biological phenomena that makes it beneficial to biologists to reason in one way rather than another, given their purposes for doing so (Lean 2023; Ross 2023; Wimsatt 2023).

5. CLASSIFICATION, NATURAL KINDS, AND PLURALISM

Scientific inquiry, as with all reasoning about the world, involves labeling and categorizing things, organizing what we observe and what we posit, into theoretical boxes. Scientific classification systems are, as Lorraine Daston (2004) puts it, "applied metaphysics" or "metaphysics in action": they make sense of our experience by positing general types of which individual things in the world are tokens. The kinds of entities and their interrelationships posited by a science are sometimes called its *ontology,* especially in the context of the formal representation of scientific data for the purposes of digital storage and sharing (Leonelli 2010; 2016; Lean 2021).

Not coincidentally, ontology also refers to the branch of metaphysics concerned with what sorts of things exist and how they are related. For a scientific metaphysician, committed as they are to basing their metaphysical theories in science, questions about ontology—about what kinds of things exist and how they fit together—can be answered by looking at how successful sciences see fit to categorize the world. From a theory-first point of view, such as that of Quine (1948), basing ontology in science means reading off the ontological commitments that scientific theories make; that is, what sorts of entity are posited, what sorts of properties they are taken to have, how these entities and their properties relate to each other, and so on. By contrast, a practice-based approach will base its ontology on classificatory schemes that organize and direct activities, techniques, and collection practices of scientists (Hacking 1983).

Whether we base ontology on theory or practice, one thing scientific metaphysicians quickly confront is that scientific pluralism is very much the norm in scientific thinking (Dupré 1995; Mitchell 2003; Kellert et al. 2006; Havstad 2018). Scientific pluralism with respect to classification is when there are multiple ways of categorizing things. Scientific pluralism can occur when different fields operate with quite different classification schemes or when one and the same field employs inherently different classification schemes (sometimes even when focusing on one and the same phenomena—see Havstad 2016). This is particularly true in the biological sciences, where widely used kind terms like "gene" and "species" appear to have multiple, often-incompatible definitions, both between and even within scientific fields (Baxter 2023; Weber 2023). There is, it seems, an apparent tension between the fact of scientific plurality and the intuition that science aims to discover the natural kinds, since the kinds posited in different sciences don't appear to fit neatly together and thus don't appear to consistently apply to a single reality (Chakravartty 2017).

We can understand this problem as that of the relationship between scientific ontologies, as ways of conceptualizing a domain for empirical purposes, and ontology as a metaphysical inquiry: In what sense, if at all, do the kinds posited by the sciences reflect kinds that actually exist in nature? More specifically, how, if at all, can we reconcile the plurality of scientific ontologies with some kind of unity in the world they supposedly represent? Is the unity of science something worth striving for? If so, what kind of unity, and what for?

These questions potentially have concrete practical implications, as they connect in important ways to our ideas about what science is for and how it should be evaluated and therefore to socially important issues such as policy-making (Mitchell 2003). This is particularly evident in the context of data-centric biology: the increasing emphasis on sharing and reuse of data across different research contexts creates the need for some form of agreement about data representation and labeling, which has raised important questions about the theoretical assumptions underlying data classification systems and how these differ between different disciplines and fields. At stake here are important issues about how to achieve cooperation and coordination across research communities while also respecting dissensus and the differing needs and goals of those communities (Smith et al. 2007; Leonelli 2010; 2016, Laubichler et al. 2018; Sterner et al. 2020; Lean 2021).

A way to address this problem, we believe, lies in carefully considering the way in which the varying purposes for which we categorize the world interact with the features of the world on which the satisfaction of those purposes depends. As Ereshefsky and Reydon (2023) point out, while categorization systems in science are to be evaluated by how well they serve their particular function, it is often, at least partly, the world being categorized that underwrites that success. Nevertheless, the function or purpose of that categorization is as inextricable to that justification as the aspects of the world that ground their satisfaction. By analogy, suppose we discovered a toolbox and wanted to justify why it contains that specific set of tools. No one tool can do everything, and so we need several. Yet a worker shouldn't be overladen, so one should also avoid unnecessary redundancy. There are tradeoffs to be made in this respect, but reasons can be given for the choices made. Understanding a toolkit requires reference to the sorts of objects and materials the tools are designed to work on but also to the user(s) of the tools and their aims, needs, and abilities. When we look at scientific classifications in this way, there is no conflict between scientific pluralism and the possibility of scientific metaphysics, or vice versa: the fact that we need multiple tools does not threaten the idea of a single shared reality, nor does the unity of reality demand that we throw away all but one of our tools. In short, another approach to unifying the sciences, besides relating them all to a theoretical hub of fundamental physics, is to relate them all to scientific practice—to conceive them as means for achieving a variety of ends and setting about understanding those means-ends relationships.

6. CHAPTER SUMMARIES

WIMSATT. In "Evolution and the Metabolism of Error: Biological Practice as Foundation for a Scientific Metaphysics," Bill Wimsatt offers a vision of scientific metaphysics based on two ideas: First, science as a cultural activity evolves through human organisms interacting with their environment. This engagement with the world makes a realist metaphysics possible. Second, scientists have developed useful heuristics for investigating problems similar to those faced in ontology and epistemology. Co-opting their heuristics makes learning about the metaphysics of the world easier.

Wimsatt has long championed the use of the concept of robustness as a central concept to evolution and science, and he here extends this further into philosophy. As he defines it, "Robustness is the use of multiple independent means of less than perfect reliability (often far less) to secure a net reliability higher (and often under a wider range of conditions) than possible with any single method." Robustness has ontological, epistemological, and methodological significance. The basic ontological claim concerning robustness states that real objects and properties are independently detectable. This generalizes the idea that the primary qualities of objects are those accessible from multiple sensory modalities. Robustness also ties the ontological status to the epistemic means of access and implies that we know when objects and properties are real when we can detect them with independent means.

Methodologically, Wimsatt contrasts analyzing scientific reasoning in terms of robustness with analyzing it in terms of deductively valid arguments. Robustness then leads him to lay out a new program for the philosophical analysis of science based on eleven heuristics used successfully in the sciences to learn about the world. The central heuristics are the first two: "Instead of looking for inexorable arguments, we look for robust tendencies; and for conditions under which those tendencies are more likely to be realized." This is how we analyze what and why we know what we know and learn where and when we are likely to know or learn more. "Instead of looking for truths, we study errors, and how they are made." Studying errors is the primary means by which we improve our tools and methods.

Wimsatt then gives six properties of heuristics which link together scientific heuristics with biological adaptations as evolutionary products. For example, these heuristics, like biological adaptations, are more "cost-effective" than infallible solutions to problems. These properties of heuristics and adaptations in part give reasons why the methodology based on robustness outlined previously is applicable to philosophy.

BAUSMAN. In "How to Infer Metaphysics from Scientific Practice as a Biologist Might," William Bausman addresses perhaps the central methodological question of this volume: How do we make well-supported inferences from biological practice to metaphysics? The basis of his answer is that the relationship between a practice and the world is *fit for a purpose.* He then pursues the following analogy: successful practices are adapted to their domain in the same way that successful organisms are adapted to their environment. The fruit of this analogy is that metaphysicians can co-opt the methodology with which biologists use the traits of organisms as environmental proxies on the basis of adaptation. Painstakingly reconstructing a contemporary research program using the height of fossil mammalian teeth as a proxy for past environmental water level, he draws general lessons for establishing traits as environmental proxies on the basis of adaptation. For instance, it needs to be shown that the increased height of teeth in grazing mammalian teeth is correlated with, functionally adaptive to, and actually an adaptation to decreased environmental water levels.

Bausman then applies the lessons learned from biological practice to scientific metaphysics. Ken Waters's (2017) argument for the no general structure thesis is a proving ground for whether metaphysicians can follow the biologists' methodology. Several problems emerge. For instance, how can we establish a correlation between a feature of practice and the metaphysics of the world? Solving this access problem seems to require independent access to the structure of the world, which we lack. Following this, how can scientific metaphysicians support the idea that a feature of practice is functionally adaptive to some part of the world? Biological practice provides promising suggestions for how to move forward on these problems, but the prospects for surmounting them are daunting. He concludes by reflecting on the relationship between metaphysics and science. Penelope Maddy (2007) distinguishes between one-tier views where metaphysical questions are continuous with scientific practice and two-tier views where metaphysical questions float freely above science. Bausman argues that his proposal does not fit neatly into either category, aiming to scientifically answer metaphysical questions that do not directly affect scientific practice.

CREATH. It is no secret that the relationship between science and metaphysics, since they became distinct disciplines, has been unclear and often strained. In "What Was Carnap Rejecting When He Rejected Metaphysics?," Richard Creath considers an episode of this history through the eyes of Rudolf Carnap. As a figurehead of the school of logical empiricism, Carnap is notorious for his rejection of metaphysics as inimical to scientific progress. Yet since "metaphysics" means different things to different people, there is much misunderstanding of precisely what Carnap was rejecting. Creath sheds light on this question by looking at how Carnap viewed different "metaphysical" pursuits. Since he was sympathetic to some of these, and indeed engaged in them himself, we should understand what he took metaphysics to be based on the works that he explicitly criticized as such. Creath's central example is French philosopher Henri Bergson; specifically, Bergson's criticism of Einstein's theory of relativity. In short, the metaphysics of Bergson that Carnap rejected was one that takes philosophical intuition to override science on the basis that the former reaches deeper into the intrinsic nature of reality than the mere abstract symbols of scientific reasoning. Carnap rejects this metaphysics because it merely compounds disagreements rather than resolving them, since incompatible philosophies can all appeal to intuition with no way to arbitrate between them. Yet there are ways of understanding metaphysics that give it a compatible and important role in science. A Carnapian metaphysics can be one that offers pragmatic recommendations for how to organize our experience of the world. Further, Creath's discussion of Carnap opens up the exciting possibility of a metaphysics on par with the sciences that produces truth-apt, empirical claims. Creath's discussion helps set the tone for the volume by clarifying how metaphysics can coexist harmoniously with a focus on scientific practice and empirical success.

LEAN. In "Ideal Observations: Information and Causation in Biological Practice," Oliver Lean discusses the relationship between two apparently quite different ways of thinking about biological systems—namely, *causality* and *information*. Like other sciences, biologists regularly discuss biological phenomena in terms of causes and effects, processes and mechanisms. Yet along with this, it is also common to hear those phenomena described in terms of information. Information is said to be carried by genes, gathered from the environment, processed by brains, transmitted in animal signals, and so on. The question of the relationship between these two ways of describing biological phenomena reveals at its core a metaphysical puzzle descended from older controversies about the relationship between mind and world: information is sometimes thought to "float free" of its physical "substrate" in some sense while at the same time being dependent on it. Opinion among philosophers and scientists about the role and value of information-talk is sharply divided. For some, it is a useful metaphor for what are "really" just ordinary causal-physical phenomena. For others, the processing of information is precisely what *distinguishes* biological phenomena from the nonliving physical world. For others, information is metaphysically suspect and scientifically misleading and should be dispensed with in favor of more naturalistically respectable talk of physical stuff and their causes.

Lean proposes a way to resolve this controversy that takes a practice-focused approach. As discussed elsewhere in the volume, causality can be very fruitfully understood in connection with practices related to manipulation and control: discovering the causal structure of a phenomenon is to discover the ways it can be changed by intervention. Lean proposes that information can be understood in a similarly practice-focused way. However, this does not mean recasting biological information in causal language, as some have done, because information is connected to a distinct set of purposes from those of causation. That is, information-talk is appropriate when the aim is to understand how living things solve certain functional problems—problems like how to adapt behavior to one's external environment, to other organisms, or to other subsystems of the same organism. With this in mind, Lean sketches a framework for informational thinking that is analogous and complementary to Woodward's (2003; 2010) theory of causal reasoning: where Woodward characterizes causal claims in terms of control through "ideal interventions," Lean characterizes informational claims in terms of coordination through *ideal observations*. Causality and information are related by the systems of practices to which each is connected, which are importantly distinct but intricately related to each other.

KAISER AND TRAPPES. Mechanisms are a pervasive structure in the biological world and for this reason have received much attention from philosophers of science. Philosophical attention has focused on molecular mechanisms (such as protein synthesis, gene expression, and neuronal transmission) and mechanisms of natural selection and ecology. However, in "Individual-Level Mechanisms in Ecology and Evolution" Marie Kaiser and Rose Trappes wish to characterize and defend another set of mechanisms under investigation by a group of behavioral and evolutionary ecologists at the Collaboratory Research Centre (CRC). This set of mechanisms are distinct from the cases previous philosophers have studied in that they operate at the individual organism level instead of at population or sub-organismal levels. "Individual-level mechanisms" involve a focal individual whose behaviors and interactions with their biotic and abiotic environment bring about a match between the individual's phenotype and its environment. The character an individual's interaction takes determines whether the mechanism is one of three types—niche construction, niche choice, or niche conformance. Collectively, this set of mechanisms are referred to as NC^3 mechanisms.

Kaiser and Trappes employ a novel methodology in their study of NC^3 mechanisms. As collaborators on the CRC project, Kaiser and Trappes have unique access to the epistemic practices of the scientists in the research group. They employ a variety of empirical approaches to describe and interpret NC^3 mechanisms as the researchers understand them—such as analysis of research grants, lectures, and publications as well as qualitative interviews and questionnaires. Their findings demonstrate a consensus (developed over the course of the research program) on the use and meaning of the concept "mechanism" by investigators. Investigators justify their adoption of a mechanism concept by appealing to causal interactions, processes, and complex organization as important ingredients for explaining how specific outcomes obtain. Kaiser and Trappes point out that these reasons happen to accord well with how the New Mechanists understand the concept. They argue that their analysis of NC^3 mechanisms involves formulating tentative metaphysical claims about the ontology at work in the CRC research program. Claims about NC^3 mechanisms are claims about what the world is like, which kinds of entities exist, and the nature of such entities. They implicitly recognize the further justification required to assert metaphysical claims in the sense we described previously as scientific metaphysics.

BAXTER. Scientific pluralism is a thesis that has been increasingly defended by philosophers of science. Roughly, scientific pluralism is the view that the world scientists investigate is characterized by a number of local, irreducible frameworks that feature incommensurable (in some sense) ontologies. Scientific pluralism is a thesis that has been advanced by philosophers of science concerning gene concepts in classical and molecular biology (Waters 1994; 2004; Griffiths and Stotz 2006; Weber 2023). Scientific pluralism prompts a variety of questions about the units of philosophical analysis. In "Just How Messy Is the World?," Baxter investigates how many distinct frameworks are under investigation in contemporary molecular genetics and the scope of explanatory and investigative significance for any given framework. She argues that the pluralism of gene concepts in contemporary molecular biology is more radical than what has thus far been characterized by philosophers of science.

Baxter's analysis draws on the individuation and annotation practices of database curators to describe and interpret the GenBank gene concept at work in the National Center for Biotechnology's digital database. In doing so, she demonstrates that genomicists employ a capacious concept that counts nucleic acid sequences as molecular genes that have been omitted from other molecular gene concepts (such as Waters 1994 and Griffiths and Stotz 2006). She argues that philosophers of science have mischaracterized scientific pluralism in genetics in two ways. First, they have undercounted the number of incommensurable gene concepts at work in contemporary molecular biology. The number of entities geneticists consider as molecular genes is actually greater than what philosophers of science have acknowledged. Second, philosophers of science have also mischaracterized the explanatory and investigative scope of molecular genes. The GenBank gene concept individuates molecular genes according to their phenotypic effects—not merely by the biomolecules whose information they encode.

Provided one is justified in inferring something about the structure of the world from the ontological commitments of geneticists—as Waters (2017) presupposes—then Baxter's discussion shows that the world is even messier than what philosophers like Waters have appreciated.

WEBER. What happens to an older science when a newer science emerges and appears to supersede the former, as in the case of, say, Newtonian and quantum mechanics? A common temptation among philosophers and scientists has been to say that the newer science has reduced the older one. Philosophers have provided various models for scientific reduction. However, in "The Reduction of Classical Experimental Embryology to Molecular Developmental Biology: A Tale of Three Sciences," Weber argues that existing accounts of reduction in philosophy simply cannot account for the relationship between classical experimental embryology and molecular biology. Classical embryology and molecular biology don't feature many (if any) fundamental laws with broad explanatory scope. This makes the case of classical embryology and molecular biology hard to accommodate by models of reduction. Instead, Weber concludes that Jaegwon Kim's (2007) account holds the most promise of accounting for the case of classical embryology and molecular biology. Weber examines the history of the organizer concept of classical embryology to demonstrate that this concept's relationship to molecular biology cannot be characterized as one of reduction in Kim's sense.

More importantly, Weber contends it was hardly the theoretical insights that the organizer concept facilitated that account for its success. The organizer concept introduced by Hans Spemann and Hilde Mangold in the early twentieth century emerged from experimental work whereby they transplanted a bit of embryonic tissue from one species of newt to another. The transplant appeared to induce the growth of another body. According to Weber, the explanatory and causal significance of the organizer concept was, at best, modest. There was much speculation among scientists with little progress over the physical realizers of the organizer concept. Nevertheless, the organizer concept is crucial to developmental biology. Why might this be? Weber proposes that what accounts for the success of the organizer concept are the set of experimental practices used to investigate developmental processes. Although the organizer concept is not reducible to the molecular level, its experimental practices continue to live on in some form in contemporary molecular biology. Even though the methods of classical experimental embryology operate at a higher level than molecular biology—for example, the cellular—methods like tissue transplantation have been integrated with contemporary methods that operate at the lower, molecular level.

Weber's discussion leads us to a very different picture of what happened with the transition from classical embryology to molecular biology than what Kim's model suggests. Instead of reduction, we have what Weber calls *inter-level investigative practice*. This framework characterizes the combination and integration of experimental interventions of an older science that occur at one level (or set of levels)—say, the whole organism, embryonic tissues, or cells—with more contemporary experimental methods that occur at different levels. Weber's idea of inter-level investigative practice represents a substantial shift in the way philosophers should think about the transition from one science to another.

ROSS. In "Explanation in Contexts of Causal Complexity: Lessons from Psychiatric Genetics," Lauren Ross discusses efforts to uncover the genetic causes of psychiatric illness. Many argue that psychiatry is in a state of crisis or immaturity: unlike somatic ailments, little progress appears to have been made in understanding the genetic causes of psychiatric disease. According to some, this shows that the genetic level is simply the wrong level at which to understand psychiatric disorders. Ross offers a new diagnosis of the controversy. First, the apparent immaturity of psychiatric genetics is at least partly because of the nature of its subject. Psychiatric illnesses are highly complex in two ways: single instances of disease tend to have multiple genetic contributors, and the causes of a disease vary widely from one instance to another. Second, despite this, scientists are finding ways to overcome the challenges that their domain presents to them. As scientists work to make their practices more fit and useful, it is important for the philosopher to consider both how the world constrains and frustrates the efforts of scientists to investigate it and what we can learn about the world from looking at practices. Ross's discussion is a rich example of how the peculiar complexity of biological systems present challenges for scientific attempts to explain them—even to classify and categorize them in the first place—and the strategies scientists use to surmount these challenges. This perspective is pragmatic, a matter of fitting one's investigative strategy to the nature of the thing being investigated given one's purposes for doing so and shows that different strategies may be needed to fit different kinds of causal complexity.

ERESHEFSKY AND REYDON. In "The Grounded Functionality Account of Natural Kinds," Marc Ereshefsky and Thomas Reydon offer a practice-oriented view of the problem of natural kinds. It is often said that science aims to classify things in the natural world according to how the natural world itself is classified. For example, there are facts of the matter about what species exist and biology's task is to correctly assign organisms to species according to those facts. Philosophers have offered several theories of what these natural kinds are that science supposedly aims to discover and represent in its classification systems. Ereshefsky and Reydon argue that existing theories fail in at least one of two ways. Some hinge on a priori matters such as whether natural kinds are universals and what kind of universals they are. These accounts fail, the authors argue, because they offer no clues about how and why scientists choose certain classification systems over others; that is, for what makes certain classification systems successful. Other theories rightly aim to account for the scientific uses of classification systems of scientific aims such as prediction but fail to account for the diversity of purposes for which scientists develop classifications. Ereshefsky and Reydon propose a grounded functionality account (GFA) of natural kinds that solves these problems. On this account, a classification scheme is to be judged solely by how well it serves the aims (epistemic or otherwise) for which it is designed. To avoid admitting too many things, such as arbitrary conventions or pure social constructs, there is a further criterion: Crucially, the kinds in the classification are *natural* kinds when the serving of those aims depends at least partly on some feature of the world that the classification system is about. The authors argue this account sidesteps the growing skepticism about the idea of natural kinds. They identify a broad agreement among biologists that the difference between natural and nonnatural is important. That is, our classification systems should in some way be grounded in the world or, rather, in the language of the GFA. The disagreements that arise are simply about which aspects of the world should ground their classifications.

REFERENCES

Ankeny, R. A., and S. Leonelli. 2016. "Repertoires: A Post-Kuhnian Perspective on Scientific Change and Collaborative Research." *Studies in History and Philosophy of Science Part A* 60: 18–28.

Baird, D. 2004. *Thing Knowledge: A Philosophy of Scientific Instruments.* Berkeley: University of California Press.

Bausman, W. C. 2023. "How to Infer Metaphysics from Scientific Practice as a Biologist Might." In *From Biological Practice to Scientific Metaphysics,* edited by W. C. Bausman, J. K. Baxter, and O. M. Lean. Minneapolis: University of Minnesota Press.

Baxter, J. K. 2023. "Just How Messy Is the World?" In *From Biological Practice to Scientific Metaphysics,* edited by W. C. Bausman, J. K. Baxter, and O. M. Lean. Minneapolis: University of Minnesota Press.

Bealer, G. 1982. *Quality and Concept.* Oxford: Oxford University Press.

Bennett, K. 2016. "There Is No Special Problem with Metaphysics." *Philosophical Studies* 173, no. 1: 21–37.

Bryant, A. 2017. "Keep the Chickens Cooped: The Epistemic Inadequacy of Free Range Metaphysics." *Synthese* 197, no. 5: 1867–87.

Carnap, R. 1931. "Überwindung der Metaphysik durch logische Analyse der Sprache." *Erkenntnis* 2: 220–41.

Carnap, R. 1950. "Empiricism, Semantics, and Ontology." *Revue Internationale de Philosophie* 4, no. 11: 20–40.

Chakravartty, A. 2010. "Perspectivism, Inconsistent Models, and Contrastive Explanation." *Studies in History and Philosophy of Science Part A* 41, no. 4: 405–12.

Chakravartty, A. 2013. "Naturalized Metaphysics." In *Scientific Metaphysics,* edited by D. Ross, J. Ladyman, and H. Kincaid, 27–50. Oxford: Oxford University Press.

Chakravartty, A. 2017. *Scientific Ontology: Integrating Naturalized Metaphysics and Voluntarist Epistemology.* Oxford: Oxford University Press.

Chang, H. 2004. *Inventing Temperature: Measurement and Scientific Progress.* Oxford: Oxford University Press.

Chang, H. 2012. *Is Water H₂O?: Evidence, Realism and Pluralism.* Dordrecht: Springer.

Chang, H. 2014. "Epistemic Activities and Systems of Practice: Units of Analysis in Philosophy of Science after the Practice Turn." In *Science after the Practice Turn in the Philosophy, History and Social Studies of Science,* edited by L. Soler, S. Zwart, M. Lynch, and V. Israel-Jost, 67–79. New York: Routledge.

Chang, H. 2017. "VI—Operational Coherence as the Source of Truth." *Proceedings of the Aristotelian Society.* Oxford: Oxford University Press.

Chirimuuta, M. 2016. "Vision, Perspectivism, and Haptic Realism." *Philosophy of Science* 83, no. 5: 746–56.

Craver, C. F. 2007. *Explaining the Brain: Mechanisms and the Mosaic Unity of Neuroscience.* Oxford: Oxford University Press.

Craver, C. F., and W. Bechtel. 2007. "Top-Down Causation without Top-Down Causes." *Biology & Philosophy* 22, no. 4: 547–63.

Creath, R. 2023. "What Was Carnap Rejecting When He Rejected Metaphysics?" In *From Biological Practice to Scientific Metaphysics,* edited by W. C. Bausman, J. K. Baxter, and O. M. Lean. Minneapolis: University of Minnesota Press.

Currie, A. 2018. *Rock, Bone, and Ruin: An Optimist's Guide to the Historical Sciences.* Cambridge, Mass.: MIT Press.

Daston, L. 2004. "Type Specimens and Scientific Memory." *Critical Inquiry* 31, no. 1: 153–82.

Dewey, J. 1958. *Experience and Nature.* Chicago: Open Court Publishing.

Dupré, J. 1995. *The Disorder of Things: Metaphysical Foundations of the Disunity of Science.* Cambridge, Mass.: Harvard University Press.

Dupré, J. A., and D. J. Nicholson. 2018. "A Manifesto for a Processual Philosophy of Biology." In *Everything Flows: Towards a Processual Philosophy of Biology,* edited by D. J. Nicholson and J. Dupré. Oxford: Oxford University Press.

Ereshefsky, M., and T. A. C. Reydon. 2023. "The Grounded Functionality Account of Natural Kinds." In *From Biological Practice to Scientific Metaphysics,* edited by W. C. Bausman, J. K. Baxter, and O. M. Lean. Minneapolis: University of Minnesota Press.

Gannett, L. 1999. "What's in a Cause? The Pragmatic Dimensions of Genetic Explanations." *Biology and Philosophy* 14, no. 3: 349–73.

Giere, R. N. 2010. *Scientific Perspectivism.* Chicago: University of Chicago Press.

Gillet, C. 2021. "Using Compositional Explanations to Understand Compositional Levels: Why Scientists Talk about 'Levels' in Human Physiology." In *Levels of Organization in the Biological Sciences,* edited by D. Brooks, J. DiFrisco, and W. C. Wimsatt. Cambridge, Mass.: MIT Press.

Glennan, S. 2017. *The New Mechanical Philosophy.* Oxford: Oxford University Press.

Green, S., and R. Batterman. 2017. "Biology Meets Physics: Reductionism and Multi-Scale Modeling of Morphogenesis." *Studies in History and Philosophy of Science Part C: Studies in History and Philosophy of Biological and Biomedical Sciences* 61: 20–34.

Griffiths, P., E. Machery, and S. Linquist. 2009. "The Vernacular Concept of Innateness." *Mind & Language* 24, no. 5: 605–30.

Griffiths, P., and K. Stotz. 2013. *Genetics and Philosophy: An Introduction.* Cambridge: Cambridge University Press.

Griffiths, P. E., and K. Stotz. 2006. "Genes in the Postgenomic Era." *Theoretical Medicine and Bioethics* 27, no. 6: 499.

Hacking, I. 1983. *Representing and Intervening: Introductory Topics in the Philosophy of Natural Science.* Cambridge: Cambridge University Press.

Haslanger, S. 2000. "Gender and Race: (What) Are They? (What) Do We Want Them to Be?" *Noûs* 34, no. 1: 31–55.

Haslanger, S. 2016. "What Is a (Social) Structural Explanation?" *Philosophical Studies* 173, no. 1: 113–30.

Havstad, J. C. 2016. "Protein Tokens, Types, and Taxa." In *Natural Kinds and Classification in Scientific Practice,* edited by C. Kendig, 94–106. New York: Routledge.

Havstad, J. C. 2018. "Messy Chemical Kinds." *The British Journal for the Philosophy of Science* 69, no. 3: 719–43.

Hawley, K. 2006. "Science as a Guide to Metaphysics?" *Synthese* 149, no. 3: 451–70.

Hempel, C. G. 1945. "Studies in the Logic of Confirmation (I.)." *Mind* 54, no. 213: 1–26.

Hempel, C. G., and P. Oppenheim. 1948. "Studies in the Logic of Explanation." *Philosophy of Science* 15, no. 2: 135–75.

Ismael, J. 2013. "Causation, Free Will, and Naturalism." In *Scientific Metaphysics,* edited by D. Ross, J. Ladyman, and H. Kincaid, 208–35. Oxford: Oxford University Press.

Ismael, J. 2017. "An Empiricist's Guide to Objective Modality." In *Metaphysics and the Philosophy of Science: New Essays,* edited by M. Slater and Z. Yudell, 109–25. Oxford: Oxford University Press.

Kaiser, M. I. 2015. *Reductive Explanation in the Biological Sciences.* New York: Springer.

Kaiser, M. I., and R. Trappes. 2023. "Individual-Level Mechanisms in Ecology and Evolution." In *From Biological Practice to Scientific Metaphysics,* edited by W. C. Bausman, J. K. Baxter, and O. M. Lean. Minneapolis: University of Minnesota Press.

Kellert, S. H., H. E. Longino, and C. K. Waters. 2006. "The Pluralist Stance." In *Scientific Pluralism,* edited by S. H. Kellert, H. E. Longino, and C. K. Waters, 19. Minneapolis: University of Minnesota Press.

Kendig, C. 2015. *Natural Kinds and Classification in Scientific Practice.* New York: Routledge.

Kim, J. 2007. *Physicalism, or Something Near Enough.* Princeton: Princeton University Press.

Kincaid, H. 2013. "Introduction: Pursuing a Naturalist Metaphysics." In *Scientific Metaphysics,* edited by D. Ross, J. Ladyman, and H. Kincaid, 1–26. Oxford: Oxford University Press.

Korb, K. B. 2004. "Introduction: Machine Learning as Philosophy of Science." *Minds and Machines* 14, no. 4: 433–40.

Kuhn, T. S. 1962. *The Structure of Scientific Revolutions.* Chicago: University of Chicago Press.

Ladyman, J., and D. Ross. 2007. *Every Thing Must Go: Metaphysics Naturalized.* Oxford: Oxford University Press.

Laubichler, M. D., S. J. Prohaska, and Peter F. Stadler. 2018. "Toward a Mechanistic Explanation of Phenotypic Evolution: The Need for a Theory of Theory Integration." *Journal of Experimental Zoology Part B: Molecular and Developmental Evolution* 330, no. 1: 5–14.

Lean, O. M. 2021. "Are Bio-ontologies Metaphysical Theories?" *Synthese* 199, no. 3: 11587–608.

Lean, O. M. 2023. "Ideal Observations: Information and Causation in Biological Practice." In *From Biological Practice to Scientific Metaphysics,* edited by W. C. Bausman, J. K. Baxter, and O. M. Lean. Minneapolis: University of Minnesota Press.

Lean, O. M., L. Rivelli, and C. H. Pence. 2021. "Digital Literature Analysis for Empirical Philosophy of Science." *The British Journal for Philosophy of Science* (March).

Leonelli, S. 2010. "Documenting the Emergence of Bio-ontologies: Or, Why Researching Bioinformatics Requires HPSSB." *History and Philosophy of the Life Sciences* 32, no. 1: 105–25.

Leonelli, S. 2016. *Data-Centric Biology: A Philosophical Study.* Chicago: University of Chicago Press.

Leonelli, S., and R. A. Ankeny. 2015. "Repertoires: How to Transform a Project into a Research Community." *BioScience* 65, no. 7: 701–8.

Lombrozo, T. 2010. Causal–Explanatory Pluralism: How Intentions, Functions, and Mechanisms Influence Causal Ascriptions. *Cognitive Psychology* 61, no. 4: 303–32.

Longino, H. E. 1990. *Science as Social Knowledge: Values and Objectivity in Scientific Inquiry.* Princeton: Princeton University Press.

Love, A. C. 2009. "Typology Reconfigured: From the Metaphysics of Essentialism to the Epistemology of Representation." *Acta Biotheoretica* 57, no. 1–2: 51–75.

Lowe, E. J. 2002. *A Survey of Metaphysics.* Oxford: Oxford University Press.

Machamer, P., L. Darden, and C. F. Craver. 2000. "Thinking about Mechanisms." *Philosophy of Science* 67, no. 1: 1–25.

Machery, E., S. Stich, D. Rose, A. Chatterjee, K. Karasawa, N. Struchiner, S. Sirker, N. Usui, and T. Hashimoto. 2017. "Gettier across Cultures." *Noûs* 51, no. 3: 645–64.

Maddy, P. 2007. *Second Philosophy: A Naturalistic Method.* Oxford: Oxford University Press.

Massimi, M. 2018. "Four Kinds of Perspectival Truth." *Philosophy and Phenomenological Research* 96, no. 2: 342–59.

Meincke, A. S., and J. Dupré. 2020. *Biological Identity: Perspectives from Metaphysics and the Philosophy of Biology.* Abingdon: Routledge.

Mitchell, S. D. 2003. *Biological Complexity and Integrative Pluralism.* Cambridge: Cambridge University Press.

Mizrahi, M. 2020. "Hypothesis Testing in Scientific Practice: An Empirical Study." *International Studies in the Philosophy of Science* 33, no. 1: 1–21.

Morrison, M. 2011. "One Phenomenon, Many Models: Inconsistency and Complementarity." *Studies in History and Philosophy of Science Part A* 42, no. 2: 342–51.

Mumford, S., and M. Tugby. 2013. "What Is the Metaphysics of Science?" In *Metaphysics and Science,* edited by S. Mumford and M. Tugby. Oxford: Oxford University Press.

Nersessian, N. J. 2012. *Faraday to Einstein: Constructing Meaning in Scientific Theories.* Dordrecht: Springer.

Neurath, O. 1931/1983. "Physicalism." In *Philosophical Papers 1913–1946,* edited by R. S. Cohen and M. Neurath, 48–51. Dordrecht: D. Reidel.

Novick, A. 2017. "Metaphysics and the Vera Causa Ideal: The Nun's Priest's Tale." *Erkenntnis* 82, no. 5: 1161–76.

Overton, J. A. 2013. "'Explain' in Scientific Discourse." *Synthese* 190, no. 8: 1383–405.

Paul, L. A. 2012. "Metaphysics as Modeling: The Handmaiden's Tale." *Philosophical Studies* 160, no. 1: 1–29.

Pearl, J. 2009. *Causality.* Cambridge: Cambridge University Press.

Peirce, C. S. 1878/1935. "How to Make Our Ideas Clear." In *Collected Papers of Charles Sanders Peirce, Volume V,* edited by C. Hartshorne and P. Weiss. Cambridge, Mass.: Harvard University Press.

Pence, C. H., and G. Ramsey. 2018. "How to Do Digital Philosophy of Science." *Philosophy of Science* 85, no. 5: 930–41.

Pickering, A. 1992. *Science as Practice and Culture.* Chicago: University of Chicago Press.

Quine, W. V. 1948. "On What There Is." *The Review of Metaphysics,* 2, no. 5: 21–38.

Reichenbach, H. 1938. *Experience and Prediction: An Analysis of the Foundations and the Structure of Knowledge.* Chicago: University of Chicago Press.

Reisch, G. A. 2005. *How the Cold War Transformed Philosophy of Science: To the Icy Slopes of Logic.* Cambridge, Mass.: Cambridge University Press.

Riesch, H. 2015. "Reductionism as an Identity Marker in Popular Science." In *Empirical Philosophy of Science,* edited by S. Wagenknecht, N. J. Nersessian, and H. Andersen, 83–103. Dordrecht: Springer.

Roe, N. 2015. "Scientific Metaphysics or the Metaphysics of Science?" Seminar paper. University of Calgary.

Rose, D. 2020. "The Ship of Theseus Puzzle." In *Oxford Studies in Experimental Philosophy,* edited by T. Lombrozo, J. Knobe, and S. Nichols, 3: 158. Oxford: Oxford University Press.

Ross, L. N. 2018. "Causal Selection and the Pathway Concept." *Philosophy of Science* 85, no. 4: 551–72.

Ross, L. N. 2020. "Causal Concepts in Biology: How Pathways Differ from Mechanisms and Why It Matters." *The British Journal for the Philosophy of Science.* 72, no. 1: 131–58.

Ross, L. N. 2023. "Explanation in Contexts of Causal Complexity: Lessons from Psychiatric Genetics." In *From Biological Practice to Scientific Metaphysics,* edited by W. C. Bausman, J. K. Baxter, and O. M. Lean. Minneapolis: University of Minnesota Press.

Ross, L. N. Forthcoming. "Causal Control: A Rationale for Causal Selection." In *Philosophical Perspectives on Causal Reasoning in Biology,* edited by C. K. Waters. Minneapolis: Minnesota University Press.

Rouse, J. 2002. *How Scientific Practices Matter: Reclaiming Philosophical Naturalism.* Chicago: University of Chicago Press.

Russell, B. 1913. "On the Notion of Cause." *Proceedings of the Aristotelian Society* 13: 1–26.

Simon, H. A. 1996. *The Architecture of Complexity.* Cambridge, Mass.: MIT Press.

Smith B., M. Ashburner, C. Rosse, J. Bard, W. Bug, W. Ceusters, L. J. Goldberg, et al. 2007. "The OBO Foundry: Coordinated Evolution of Ontologies to Support Biomedical Data Integration." *Nature Biotechnology* 25, no. 11: 1251–55.

Soler, L., S. Zwart, M. Lynch, and V. Israel-Jost. 2014. *Science after the Practice Turn in the Philosophy, History, and Social Studies of Science.* Abingdon: Routledge.

Spirtes, P., C. Glymore, and R. Scheines. 2000. *Causation, Prediction, and Search.* Cambridge, Mass.: MIT Press.

Sterner, B., J. Witteveen, and N. Franz. 2020. "Coordinating Dissent as an Alternative to Consensus Classification: Insights from Systematics for Bioontologies." *History and Philosophy of the Life Sciences* 42, no. 1: 8.

Sullivan, M., and P. Van Inwagen. 2020. "Metaphysics." In *The Stanford Encyclopedia of Philosophy,* edited by E. N. Zalta, Metaphysics Research Lab, Stanford University.

Van Fraassen, B. C. 1980. *The Scientific Image.* Oxford: Oxford University Press.

Waters, C. K. 1994. "Genes Made Molecular." *Philosophy of Science* 61, no. 2: 163–85.

Waters, C. K. 2004. "What Was Classical Genetics?" *Studies in History and Philosophy of Science Part A* 35, no. 4: 783–809.

Waters, C. K. 2007. "Causes That Make a Difference." *The Journal of Philosophy* 104, no. 11: 551–79.

Waters, C. K. 2008. "Beyond Theoretical Reduction and Layer-Cake Antireduction: How DNA Retooled Genetics and Transformed Biological Practice." In

The Oxford Handbook of Philosophy of Biology, edited by M. Ruse. Oxford: Oxford University Press.

Waters, C. K. 2014. "Shifting Attention from Theory to Practice in Philosophy of Biology." In *New Directions in the Philosophy of Science,* edited by M. C. Galavotti, D. Dieks, W. J. Gonzalez, et al., 121–39. Berlin: Springer International Publishing.

Waters, C. K. 2017. "No General Structure." In *Metaphysics and the Philosophy of Science: New Essays,* edited by M. Slater and Z. Yudell. New York: Oxford University Press.

Weber, M. 2006. "The Central Dogma as a Thesis of Causal Specificity." *History and Philosophy of the Life Sciences* 28, no. 4: 595–609.

Weber, M. 2023. "The Reduction of Classical Experimental Embryology to Molecular Developmental Biology: A Tale of Three Sciences." In *From Biological Practice to Scientific Metaphysics,* edited by W. C. Bausman, J. K. Baxter, and O. M. Lean. Minneapolis: University of Minnesota Press.

Wimsatt, W. C. 1976. "Reductionism, Levels of Organization, and the Mind-Body Problem." In *Consciousness and the Brain,* edited by G. Globus, G. Maxwell, and I. Savodnik, 205–67. New York: Springer.

Wimsatt, W. C. 1994. "The Ontology of Complex Systems: Levels of Organization, Perspectives, and Causal Thickets." *Canadian Journal of Philosophy* 24, no. S1: 207–74.

Wimsatt, W. C. 2007. *Re-engineering Philosophy for Limited Beings: Piecewise Approximations to Reality.* Cambridge, Mass.: Harvard University Press.

Wimsatt, W. C. 2023. "Evolution and the Metabolism of Error: Biological Practice as Foundation for a Scientific Metaphysics." In *From Biological Practice to Scientific Metaphysics,* edited by W. C. Bausman, J. K. Baxter, and O. M. Lean. Minneapolis: University of Minnesota Press.

Wolpert, D. H., and W. G. Macready. 1995. "No Free Lunch Theorems for Search." Technical Report SFI-TR-95-02-010, Santa Fe Institute.

Woodward, J. 2003. *Making Things Happen: A Theory of Causal Explanation.* Oxford: Oxford University Press.

Woodward, J. 2008. "Mental Causation and Neural Mechanisms." In *Being Reduced: New Essays on Reduction, Explanation, and Causation,* edited by J. Hohwy and J. Kallestrup, 218–62. Oxford: Oxford University Press.

Woodward, J. 2010. "Causation in Biology: Stability, Specificity, and the Choice of Levels of Explanation." *Biology & Philosophy* 25, no. 3: 287–318.

Woodward, J. 2014. "A Functional Account of Causation; Or, a Defense of the Legitimacy of Causal Thinking by Reference to the Only Standard That

Matters—Usefulness (as Opposed to Metaphysics or Agreement with Intuitive Judgment)." *Philosophy of Science* 81, no. 5: 691–713.

Woodward, J. 2015. "Methodology, Ontology, and Interventionism." *Synthese* 192, no. 11: 3577–99.

Woodward, J. 2017. "Physical Modality, Laws, and Counterfactuals." *Synthese* 197, no. 5: 1907–29.

Woody, A. I. 2015. "Re-orienting Discussions of Scientific Explanation: A Functional Perspective." *Studies in History and Philosophy of Science Part A* 52: 79–87.

EVOLUTION AND THE METABOLISM OF ERROR
Biological Practice as Foundation for a Scientific Metaphysics

WILLIAM C. WIMSATT

1. EVOLUTIONARY FRAMEWORK

The idea that biological practice should provide a foundation for a scientific metaphysics may seem bizarre. Prior attempts to construct a scientific metaphysics have had different aims and used different methodologies, seeking deductive derivations of upper-level objects, concepts, and laws from fundamental entities and relationships, usually from the lowest-level (smallest) entities and relationships of contemporary physical theory, and sought accounts according to which the upper-level things were "nothing more" than these lowest-level posits (e.g., Steven Weinberg). The other strategy is to begin from phenomenological posits, like sense data, which were seen as immediate experience of which one had direct and certain access, out of which one could deductively construct higher-level entities (objects) and relations. The first of these was an attempt at an ontological reduction, the second an epistemological one. In either case, the aim was to justify scientific theory and practice as certain and error-free (e.g., Bertrand Russell or Bas van Frassen). The path I take is a third, which recognizes the fundamental fallibility and error tolerance of our adaptive information-gathering interactions with the world. It seeks ways to ground our natures and processes that are neither deductive nor eliminative but fundamentally related to the nature of evolutionary processes and ways of producing order in the world that are intrinsically heuristic, error-tolerant, and consistent with actual adaptive constructive processes. This ultimately necessitates not only a metaphysics based on biological practice but also a correspondingly based epistemology and methodology. I aim to provide all three.

2. AN ATTEMPTED BLACKBALL FOR ANY POSSIBLE SCIENTIFICALLY BASED METAPHYSICS

When I was an undergraduate, there was an analytically based argument against the appropriateness of any possible scientifically based metaphysics. This was to note that all scientific theories were empirically based and fallible. And not only that, given the historical record, any current or new theory was overwhelmingly likely to be false. As such, it was argued, scientific theories were categorically the wrong kind of thing on which to base a metaphysical system. But this is too quick. One can accept their fallibility and even their falsehood without being ready to accept that new theoretical revisions would totally overturn our expectations about the behavior of nature. I would argue that indeed, in main outline, our macroscopic regularities would be preserved through any possible scientific revolutions. Nothing, for example, could overthrow the main results of evolutionary theory or classical mechanics. This is true for two distinct reasons: first, most macroscopic regularities are at least approximately true "sloppy, gappy causal regularities" in the sense to be expounded later, and these are robust. This means that they are not dependent on fine details of the underlying processes: changes in these processes, or in our accounts of them, do not change the macroscopic regularities. (See Batterman 2021 on multiscale explanations, generalizing, and better founding an argument I first made in 1981.) As Cartwright (1983) argued, the ideal gas law is safe in any future scientific revolution. The second reason is that it is an adequacy condition for any new theory that it be able to capture the main successes of the older theory that it replaces, most commonly as limiting cases. I claim that the elements I draw from biological practice are both robust and sufficiently central to be preserved in any future scientific revolutions. They are thus, in those respects, appropriate bases for a scientific metaphysics.

The idea that biological practice should provide a foundation for a scientific metaphysics has at least two anchor points. The first is that evolution itself as a process provides constructive principles that are exemplified in our normal and scientific activities because of the unavoidable fact that we are biological organisms and pursue our activities in the natural world that conditions physical processes including biological evolution. As Donald

Campbell (1974) said, "Every case of fit between a system and its environment is a product of selection." This fact alone argues for at least a basic recognition of the demands of the real world and a minimal scientific realism. It also provides a basic architecture for generative systems, including deduction, as extremely productive heuristics, as we shall see later. The second is that the investigative approaches used in science and manifested particularly clearly in biology point naturally to ontological, epistemological, and methodological approaches and conclusions. These insights may have emerged originally in biology but are no less applicable in the physical sciences once we know how and where to look.

My own work, beginning early in my career with my dissertation on functional organization (Wimsatt 1971; 1972) and paper on complexity and organization (1974) drawn from it, exemplifies this and provided a foundation for the functional analyses and complexities of scientific practice (including levels of organization and the role of mechanism) that I have pursued since (Wimsatt 1976a; 1976b; 2007; 2021b). Because practice is an end-directed activity, this analysis of function provides a framework for all of our activity and directly informs all of my other work. Thus, I take a teleological worldview of scientific and conceptual activity and return in a way—despite the substantial differences—to an Aristotelian worldview.[1] I must confess that I never (or rarely) thought of myself as a metaphysician until I was confronted with my own texts by Ken Waters and Alan Love, who reminded me that I was after all making metaphysical claims, and quite substantial ones. Thus, I join happily with the metaphysicians in this volume, although I recognize that traditional metaphysicians might find us a strange group. But it is the aim of this volume in part to convince them, through our efforts and products, that our memberships are fully paid up.

Robustness is usually viewed primarily as an epistemological tool, but it reveals as real objects other than those picked out by traditional metaphysicians, supporting a recognition of the "ontology of the tropical rainforest" rather than the traditional foundational minimalist and eliminative commitment to Quine's "desert ontology." So, as I argue later, it reveals an ontology as well and leads to a different way of proceeding than is common among traditional metaphysicians and epistemologists. It is also particularly interesting as a widespread architectural feature of biological organization for most more important biological functions, but the reach of this architecture extends beyond into all scientific domains, both theoretical and experimental and, as we will see, including mathematics.

3. LAWS VERSUS SLOPPY, GAPPY CAUSAL GENERALIZATIONS

Attention to evolutionary processes leads to another break with tradition in foundationalist scientific metaphysics, where exceptionless laws build the architecture of nature and the assumption is that only exceptionless regularities can be causal. The importance of "sloppy, gappy generalizations" or "sloppy, gappy causal regularities" rather than exact exceptionless laws in the compositional sciences and in engineering and technology is itself foundational[2] and due most immediately to the fact that the compositional sciences deal with mechanisms, which have tolerance levels, lifetimes, and failure rates and operate in a variety of contexts.[3] They are ultimately a product of the fact that selection in both the natural and human sciences and in the domain of technology operates most commonly by incremental differential improvement relative to the current state, so that the criteria for something to be selected are: "Is it better than what we are now using?" and, ultimately, "Is it good enough often enough?" This leads causally not only to approximate solutions but also, for material things, to the need to deal with natural populational variability that frustrates attempts at universal exceptionless generalization.

Recognizing the central role of populational variability leads us to recognize that although conceptually, causation requires that exactly the same cause (exactly the same state specification) in exactly the same circumstances will produce exactly the same effect, this formulation, while true, is irrelevant, since it is never realized in nature.[4] What is required is a formulation to recognize the kind of variation and qualifications we must deal with in the real world—a kind of statistical robustness: *quite similar causes in quite similar circumstances will sufficiently often produce quite similar effects.*

The introduction of variability within tolerances ("sloppy"), design limitations, and failure rates (exceptions or "gaps") is most obvious in technology, where design processes, confronted with the contingencies of the material world, explicitly include tolerable error rates in production and performance and designed lifetimes. But the same considerations clearly apply for biological organisms, which vary enormously in characteristic reliability and lifetimes in ways appropriate to their respective ecological niches, be it a bacterium or a baobab tree. Exceptionless laws are idealizations for real compositional systems but false or only approximately true and unrealistic for the real world. Selection goes for the satisficing solution. Like all organisms, we not only are designed to deal with error, but we utilize errors to gain information from the environment and to make more reliable structures in an ontological replay of natural selection—thus, "the metabolism of error." These "satisficing" and "fallibilist" desiderata do not articulate well with traditional foundationalist and logical paradigms, which mistake idealizations about decision making ("maximization of expected utility") and scientific inference (deduction) for analytic and foundational principles.

Another casualty of rejecting deductive foundationalism is eliminative reductionism, or "nothing-but-ism." Russell was skeptical of theoretical entities, describing them as "logical fictions," and many reductionist scientists wanted ultimately to explain everything at the lowest possible level—or, as Roger Sperry ironically put it, to explain "eventually everything in terms of essentially nothing." I have argued instead (Wimsatt 1976a; 1976b; 2007; 2021b) for a multilevel explanatory account of "articulatory reduction" in which lower- and intermediate-level interactions explain upper-level phenomena (like Brownian motion) and regularities (like the ideal gas law or the van der Waal equation of state). In this case, both the upper-level phenomena and the regularities delineating sloppy, gappy causal generalizations are explained but not eliminated by the articulation of lower-level causes. (See also Batterman 2021.)

The "Ur-principle" of the metabolism of error is profoundly heuristic and evolutionarily based and produces intrinsically heuristic robust and reliable structures that best serve the complex interactive networks of natural and human ends in our world. I will discuss the nature of heuristics after we discuss robustness.

My deepest problem with traditional philosophical views of science and of ourselves as scientific agents is thus that they adopt "in principle" idealizations of our aims, our scientific products, our data, and our methods for achieving them that are profoundly false and inappropriate for limited beings and that both flaw our attempts to truly characterize our activities and prevent us from utilizing the strengths that we can and do deploy in pursuing our aims (Wimsatt 2007). So then what resources do we have for fallible beings?

4. ROBUSTNESS

Robustness is the use of multiple independent means that may individually be of less-than-perfect reliability (often far less) to secure a net reliability higher (and often under a wider range of conditions) than possible with any single method (von Neumann 1956; Wimsatt 1981). This means that by incorporating multiple alternative and independent lines of argument to the same conclusion, we can use heuristic and inductive methods of inference as well as deductive inference in drawing reliable conclusions. Robustness can serve in the reliable and more accurate detection of objects, the construction of reliable inferences, and the production of stable results. To a deductivist, this may look like sloppy (even otiose) redundancy, but it fits our natural methods for securing reliability far better than our deductivist idealizations and allows use of our rich armamentarium of heuristic methods. Physicist Richard Feynman (1968, ch. 2) contrasted a "Babylonian" with a "Euclidian" methodology and clearly favored the former. In a "Babylonian" theory, theoretical elements were overconnected, redundant, and often approximations, and consequently results could be cross-checked and calibrated and were reliable under diverse failures. This was preferable in the real world to an elegant, sparse deductivist ("Euclidean") theory, which collapsed like a house of cards if anything failed because it had no redundancy. Our deductivism supposedly captured the structure of mature physics, though Feynman would have disagreed. Feynman is in effect arguing for a biological architecture for theories and doing so for physics as well as for biology.

But it goes deeper metaphysically. As I have argued (Wimsatt 1981), the very notion of a real object, is of something which has multiple properties, thus also has multiple means of detection and is robust. This strategy does a runaround on the traditional "argument from illusion" for the unreliability of the senses in detecting the real world by arguing that the independent means will tend to fail under different conditions and thus will reliably detect the external world when used conjointly. This is why robustness has ontological implications. (See also Eronen 2015.) And not just for objects. Within the philosophical tradition begun by Galileo and Descartes, it was the primary (multimodally detectable) qualities that were real (because they were not delimited by, and thus seen as external to, any of the senses), and the secondary qualities (detectable in only one sensory modality) were the subjective effects in us of the action of the primary qualities. That is to say, the primary qualities are robust. It is one of the ironies of the deductivist tradition from Descartes that he urges us to start with robust objects and properties in building our system but never seems to apply this lesson more broadly.

Mathematical ecologist Richard Levins (1966) urged the search for "robust theorems" that followed in multiple different independent models of a phenomenon that made diverse assumptions (Wimsatt 1981; Weisberg 2006; Odenbaugh 2011). He elsewhere spoke of this as "exploring the space of possible models" (Levins 1968). He was in effect urging the heuristic use of the criterion for logical truth as "true in all possible worlds" as a way that we could find results in the real world that we could trust. He was applying it in mathematical ecology, where all of the models used were strongly idealized approximations that were literally false. Thus, he urged that "in ecology, our truths are the intersection of independent lies" (Levins 1966). But more broadly, mathematicians in diverse areas particularly value results that show robustness through multiple independent derivational paths over those that do not (Kromer 2012; Corfield 2010; Avigad 2020), and as I mentioned earlier, Feynman (1968) argued for a similar architecture for physical theory. This is a way of preventing or massively reducing the probability of inferential failures when real people incorrectly assess the soundness or validity of an argument, and it is a principle respected and used in the most reliable of disciplines.

Robustness is the scientist's answer to the philosopher's fruitless quest for certainty. Within psychology and the social sciences, Donald Campbell (1966) sought reliability and validity through "triangulation" using multiple independent methods to calibrate and correct measurements, contrasting the heuristic that he called "multiple operationalism" with the single-linked "definitional operationalism" of philosophers and psychologists, which he regarded as fallacious. Glymour (1980) urges multiple connectedness in the localization of faults in theories (thus arguing against the Quine-Duhem hypothesis of the unfalsifiability and underdetermination of theories) and notes that a claim is untestable unless it is accessible in two or more independent ways. Robustness, though not infallible, is in effect a kind of *overdetermination* of theories. Tom Nickles gives creative extensions and qualifications to the idea of robustness in extending its use to inferential systems (Nickles 2012).

Robustness is characteristic of biological organization as well as principles of inference, as a product of selection, with redundancy, multiple means, and excess capacity to reliably accomplish all of their important functions (Wagner 2005), including their evolvability. They do so, remarkably, with stable heritability of phenotypic traits and fitness under sexual recombination (Wimsatt 1987; 2007), without which cumulative evolution would be impossible. So the same organizational design principles are found and centrally important in nature to secure reliability. So it is not surprising that scientists should follow a similar methodology in their arguments, choosing to trust entities and properties that are multiply detectable.

For most of the last century, formalist and foundationalist ideas have substantially influenced our own conceptions of what we are doing as philosophers, under the aegis of logical empiricism and analytic philosophy more broadly, but we should pay more attention to the "Babylonian" methodology of scientists. Heuristic procedures for solving problems are a species of cognitive adaptations. They are error-prone but cost-effective and applied in cross-checking ways in the complementary reduction of error. Many of the so-called "informal fallacies" of inference are prone to error but are nonetheless often quite effective and reliable. Thus, "appeal to authority" is a supposed fallacy, but it is used endemically in science, and necessarily so. No single person can know, much less validate, all of science. Campbell (1974) spoke of science as using a "95% doubt/trust ratio," in which we selectively doubted the elements under test, moving from one to another, while trusting in each case the other 95 percent—most of which we will have to take on authority or past history. This too was the lesson of Neurath's boat[5] and resonates with both biological architectures and adaptations and with real scientific methodologies. Most of the fallible inference principles that we use to increase reliability in robust arguments, such as "appeal to authority," are heuristics, so heuristics are the next topic to consider. The study of heuristics both suggests and calls for an entirely different viewpoint in constructing philosophical methodology.

5. HEURISTICS IN NATURE AND IN US

I have argued that most scientific tools of discovery, evaluation, experimental design, calibration, and theory construction are heuristic procedures.[6] I have offered a fuller analysis of these than is to be found anywhere else in the literature (Wimsatt 2007). All of the ones of which I am aware (I have primarily studied problem-solving strategies) share six properties and commonly exhibit two others. As an important marker of the connections between biological architectures and our seat-of-the pants inference procedures, with each I note an analogous or identical feature of biological adaptations. These property pairs are:

(1) By comparison with truth-preserving algorithms or other procedures for which they might be substituted, heuristics make no guarantees that they will produce a solution or the correct solution to a problem. A truth-preserving algorithm correctly applied to true premises must produce a correct conclusion. A heuristic need not.

(1′) Similarly, no adaptation guarantees success of an organism in achieving that which the adaptation is for.

(2) By comparison with procedures for which they are substituted, heuristics are "cost-effective" in terms of demands on memory, computation, or other limited resources. (This is why they are used instead of methods offering stronger guarantees.)

(2′) We take for granted that adaptations are designed in a cost-effective way—often "quick-and-dirty solutions" that work "well enough often enough."

(3) Errors produced using a heuristic are not random but systematically biased:

   (a) The heuristic will tend to break down in certain classes of cases and not others, but not at random.

   (b) With an understanding of how it works (viewing heuristics as mechanisms), it should be possible to predict the conditions under which it will fail.

   (c) Where it is meaningful to speak of a direction of error, heuristics will tend to cause errors in a certain direction, again a function of the heuristic and of the kinds of problems to which it is applied.

   (d) Heuristics may thus leave "footprints" indicating their application—results exhibiting their characteristic biases. One can work back from these to determine that a heuristic was applied and sometimes even which one.

(3′) Adaptations have a specific design that can be made to fail with a suitable choice of environmental or experimental conditions; for example, light-seeking tropism of the caterpillar of the goldtail moth is selected for because it characteristically leads the caterpillar in the spring to the top of the plant where the new leaves on which it feeds are to be found. But in a suitably designed experiment with a T-maze with the light at one end and the leaves at the other, the tropism will lead it away from rather than toward food (again, exploiting the weaknesses of the mechanism).

(4) Application of a heuristic to a problem yields a transformation of the problem into a nonequivalent but intuitively related problem that is easier to solve. Answers to the transformed problem may not be answers to the original problem, though various cognitive biases operative in learning and science may obscure this. Thus, for example, the approach to problems of heredity through transmission genetics proved much easier by ignoring the problem of how genes produced traits, and for a while transmission genetics was mistaken by some for the whole solution to the problem of heredity. (The way to completing the solution by adding an account of gene action in producing the phenotypic manifestation was begun with Jacob and Monod's elucidation of the nature and operation of the *lac* operon.)

(4′) Similarly for adaptations—for example, sensing temperature seasonality (and impending cold) by detecting changing day length.

(5) Heuristics are useful for something: they are purpose relative. Tools that are effective for one purpose may be bad for another, and increases in performance in one area are commonly accompanied by decreases elsewhere (Levins 1968). This is an instance of the "generalist" versus "specialist" trade-off. Recognizing this may help identify or predict their biases: one expects a tool to be less biased for applications it was designed for than for others it is co-opted for.

(5′) Adaptations that are clearly suboptimal because they are co-opted from a structure or procedure originally selected to serve another purpose or are subject to entrenched constraints imposed by the architecture of other adaptations are endemic in biology.

(6) Heuristics are commonly descended from other heuristics, often modified to work better in a different environment. Thus, they commonly come in related families, which may be drawn on for other resources or tools appropriate for similar tasks. Thus, Douglas Lenat (1981) exhibits some sixty heuristic rules for his theorem prover, which can also be seen as sixty instantiations of the same heuristic with slightly different antecedent conditions. Similarly one can find an array of different kinds of hammers to serve different but related purposes. On different scales of resolution, a family of heuristics may look like a single heuristic, or conversely.

(6′) Adaptations show a pattern of similarities with variations, both adaptive and not, among individuals and, at a larger scale, across species.

What does it mean that heuristics and adaptations share these six key properties? That heuristics are reducible to adaptations? No—rather that they share a common functional logic as products of evolutionary processes. But are there any other properties they might share as a result of this origin? I discuss one here, and one in the next section.

Robustness and adaptive radiations: I want to suggest that heuristics are useful in direct proportion as they are robust, even if they are both "sloppy" and "gappy" in ways indicated by the preceding six properties, so robustness is a desirable and common feature of good heuristics.

(Cartwright [1983] marks robustness as an important property of generalizations, and it is so here as well.) A heuristic that worked only under one exactly specified set of conditions and not for any of its neighbors would be of little utility, if it could be used at all. One must be careful here, since highly specialized heuristics may require very special circumstances to be useful. But when this is true, they will most commonly be a member of a larger family with a common principle that may be fruitfully specialized in different directions for different specialized uses. Thus, an adaptive radiation of heuristics is evidence of a kind of robustness of a kind of adaptation—a design principle for a class of artifacts. Thus, there are all kinds of hammers and of scissors and of threaded fasteners, each adapted to similar functions in different circumstances. (Herkimer's *Engineer's Illustrated Thesaurus* [1952] illustrates this with multiple examples in families of mechanisms.)

6. GENERATIVE ENTRENCHMENT, CONSERVATION, AND QUASI-RECAPITULATION: WHY HISTORY MATTERS

Secondly, both heuristics and adaptations, through their origin as co-opted and reselected variants of existing adaptive methods, show the marks of their history in preserved entrenched properties. This may seem merely accidental, but it is an efficient and adaptive strategy when trying to create a new tool to use what is already at hand. The co-option of existing adaptations as tools to serve new functions is central to evolutionary processes and endemic throughout, as first argued by Stephen J. Gould. It brings new selection pressures to bear on both the tool and the systematic practice or system of which it is a part. This is a major driver of new adaptive innovation and source of new complexities and opportunities.

Thus, in Wimsatt (2007), I compared nature to a "backwoods mechanic and used parts dealer," an image masterfully delineated by Douglas Harper in his *Working Knowledge: Skill and Community in a Small Shop* (1987). There the main protagonist, an able and ingenious mechanic, lived in rural upstate New York and specialized in maintaining older Saab cars. He seldom had exactly the right part in his yard of junked Saabs, but he could usually adapt (or "co-opt") and modify a similar part from another year, often in a way that improved on the intended replacement. This too uses design similarities resulting from generative entrenchment. *Generative entrenchment is an unavoidable consequence of evolution and of the evolution of increasing complexity either in biology or in cumulative culture.* In explanation, whether biological or cultural, it is the main reason that history matters.

In co-opting an element from an existing design for a new function, there may be elements, often external to the system, that aid in achieving that function. These are called scaffolds (Wimsatt and Griesemer 2007; Caporael, Griesemer, and Wimsatt 2013). A scaffold may be a behavior, an object, or a property, but in scaffolding it must exhibit an activity. A scaffold that is useful in achieving a function, if external, may subsequently be incorporated into the system for future use, increasing the organizational complexity of the design. In a complementary activity, elements of a constructed niche may be useful for other systems and become common infrastructure, tying them together into a larger ecosystem. Such processes occur with our technologies, which are highly dependent, and with ecological communities like coral reefs, where the coral backbones provide shelter and other resources for a diversity of other species.

But entrenchment alone does not determine organizational form. There may be multiple such forms, but particularly important is one whose heuristic benefits in generating easily an array of possible variations systematically have resulted in its relatively frequent occurrence and central importance in the history of evolutionary systems. These are entrenched combinatorial alphabets.

7. THE HEURISTIC ADAPTIVE ORIGINS OF COMBINATORIAL GENERATIVE SYSTEMS (INCLUDING DEDUCTION WITH FOUNDATIONAL ELEMENTS)

Elements can become fixed in their details because so much else depends on them. This is the basis for Francis Crick's famous comment that the genetic code is a "frozen accident." This I call generative entrenchment: something remains fixed or relatively stable in evolution because of its generative role; as more things come to depend upon it, the chance increases that a change in it will cause problems downstream and, if they are large enough, lethality. This decreased chance means that if other solutions exist, the waiting time to find one through random mutation is longer. This differential dependence thus leads, probabilistically, to degrees of stability with degrees of dependence and increasing stability for older elements. This feature makes cumulative evolution, both in biology and for culture and technology, possible (Wimsatt 2019). Generative entrenchment also gives biological organization a tendency to be hierarchical and fundamentally historical (Wimsatt 2015).

But another crucial property of the genetic code is its use of a small number of elements arranged in different combinations as inputs to an apparatus to assemble a correspondingly diverse array of molecular machines—linear strings of amino acids that fold into active proteins. This points to another important process and class of adaptations. As with the genetic code, entrenchment can lead to the emergence of a class of standardized parts, as with standardized thread sizes and profiles in threaded fasteners (Herkimer 1952; Wimsatt 2013). They must be standardized (within tolerances) to be interchangeable, and the tolerances commonly become more demanding as they are utilized in multiple contexts. These can themselves become "tinkertoys" *or combinatorial alphabets* in the construction of a diverse class of adaptive machines. Thus, amino acids, cells, standardized interchangeable mechanical and electronic parts, words, and program instructions are basic "alphabetical" elements capable of making large and diverse classes of machines. Some of these systems are specifically designed to facilitate reuse of parts in the evolution of complex adaptations. (Object-oriented programming, with the standardized interfaces of its objects, is also an appropriate example here. The activity there called "program maintenance" is actually more accurately called "program evolution," with the standardization of interfaces allowing the reuse of program objects in new contexts.) These are the source of major adaptive radiations and of efficient generation of "bottom-up" modular variation and increases in complexity in adaptive machines capable of occupying diverse niches.

In addition to an alphabet of stable standardized elements that can be arranged in various ways, we require for each a set of limited and constraining rules for how they may be combined—a syntax. In these, we see the basic requirements for a formal system: a set of basic elements and a set of rules for assembling them into well-formed structures, or "wffs." This is the architecture of a deductive system—and thus of all formalistic and traditional foundationalist approaches. But this arrangement is far broader. This same set of properties—basic alphabet and combination rules—is true also for atoms in making molecules, which makes "computationalism" so plausible as a form of physicalism. And it is so also for interchangeable machine parts. The stability of the standard thread specifications gives a standardized way of connecting mechanical elements. These two features are central to the generative systems mentioned previously, abstracted for the first time in the elements and propositions in Euclid's geometry and subsequently abstracted further in the development of formal symbolic logic. The possession of these two elements was the basis for Chomsky naming his theory in *Syntactic Structures* (1957) a "Generative Grammar," but if we look at Herkimer's *Engineer's Illustrated Thesaurus* (1952), we see the same thing: arrays of mechanical parts, grouped by function, mechanism, and means of articulation or connection. The last give ways of combining basic elements into a variety of diverse structures to accomplish different adaptive tasks. Function and mechanism relate to their engineering meaning, and means of articulation indicate their syntax: what can be connected to what and how.

Similar and related insights for deductive systems as adaptive structures are nicely reflected in Jeremy Avigad's (2020) analysis of the problem of reliability in the practice of the use of deductive systems in mathematics. There he notes not only the use of multiple connections or deductive paths to increase reliability and robustness but also the use of modular elements in the localization of error and the different combinatorial arrangement of these elements—an important and widespread adaptive design feature not discussed here (but see Wimsatt 2021a).

Deductive foundationalism arises from reifying the generative power and idealized error-free transmission by deductive systems of truth-like properties, starting from elements, relations, and assumptions held to be certain or true, either of basic physical properties and relations or of symbolic or logical ones, and attempting to generate or ground a much broader array of trustworthy statements, relations, and properties. These elements and the syntax used to transform them become foundations of logical inference. But taking this too seriously leads to an erroneous extrapolation of what we find to what exists in nature. This kind of constructional system is a powerful adaptation, but primarily a heuristic for dealing with the complexities in nature. In nature, nothing is certain, so we should try to ground things in robust objects, properties, or relations (like Descartes's primary qualities). As we build a system, the basic elements and transformations become successively more irreplaceable through generative entrenchment, just as the elements of the genetic code and its translation into proteins become increasingly universal and increasingly unmodifiable. But increasingly unmodifiable does not mean it is absolutely so. Any deep modifications would be very difficult, usually lethal, but when they work absolutely revolutionary. So what we need is not an absolute foundationalism but a dynamical one, one that can accomplish the ongoing repairs to Neurath's boat, however rare and difficult they may be. So a naturalistic foundationalism must be a dynamical foundationalism.

On a smaller scale, this would replicate what we see in a scientific revolution. Does this suggest a (non-progressive) Kuhnian revolution? No, it cannot be, and in fact Kuhn's views are flawed because an adequacy condition for any replacement or modification (an acceptable plank for Neurath's boat) is that it must capture the major successes of what went before. The new plank must fit (Janssen 2019; Wimsatt 2007; 2021a). It must be a functional equivalent (sometimes only approximately) for what it replaces. And now we have found the fallible, evolvable, and progressive heuristic for the generation and improvement of complex adaptive structures. The importance of deductive systems remains, but as an important heuristic methodology for problems where there is enough structure for it to be appropriate. And with this, it is seen to be no longer adequate as a traditional foundational architectonic.

8. COMBINATORIAL COMPOSITION: DEFINING, POPULATING, ANALYZING, AND USING A SPACE

Combinatorial strategies can be powerful and productive heuristics for the organized definition, population, and exploration of a space of possibilities. Definition of such a possibility space is a mark of a well-structured problem, and one of the most effective strategies in trying to bring structure to a problem that is ill-structured (Simon 1973). The Cartesian product of the defining dimensions of the space characterizes the possible elements of that space and permits systematic exploration of it and of trajectories defined by piecewise changes of the values of variables in its dimensions. Entries will have neighbors in each of these dimensions, defining their multidimensional neighborhoods, and these spaces have other useful properties. Definitions of such a space are connected with the notions of closure and completeness in logical systems. I have worked at length with two cases utilizing such spaces, one conceptual and one scientific.

The conceptual case is the characterization of types of emergence. To most scientists, unlike most philosophers, emergence and reduction are compatible, and emergence can be seen when the properties of the system are products of how the parts articulate together. But there is no direct way to characterize these different possible modes of organization. However, one can approach it negatively. Consider when the value of a system property is aggregative, so that it is nothing more than the sum of values of that property for its parts—in other words, when their organization does not matter. I delineate four requirements for aggregativity. These can be met or can fail in various ways and to various degrees in various combinations—leading to fifteen different ways in which a system property can be emergent, with aggregativity as the sixteenth case. This provides a useful combinatorial classification for different modes of dependence of a system property on the organization of its parts and tools for analysis of organization. That they are degree properties extends its usefulness by providing sensitive tools for dealing with approximations. This is discussed in Wimsatt (1997) and vastly elaborated and applied in Wimsatt (2007, ch. 12). It has rapidly become one of my most widely cited works.

Probably the best-known scientific cases are the discussions of genotype space and protein space initiated by John Maynard Smith in 1970 and widely used for a variety of conceptual and scientific arguments since. They are worth further systematic study of their use. The scientific case I have considered in detail is related: the development of the Punnett square (Wimsatt 2012) and its role in genetics through a compact and clear characterization of the possible genotypes produced in diploid matings among an array of possible gametic types. This plays a crucial role in defining matings in classical and population genetics. As discussed there, the form of algebraic expansion provided by Mendel was far less compact and intuitive, and the neat spatial organization of the Punnett square provided an easy way to reason about matings that becomes exponentially more useful and important as the number of factors considered simultaneously increases. I also discuss there why the spatial organization of the Punnett square simplified representation and problem-solving and probably was a significant factor in eliminating errors when considering multi-locus matings.

9. ROBUSTNESS, PHILOSOPHICAL METHODOLOGY, AND THE METABOLISM OF ERROR

We have elaborated the widespread use of heuristics in evolved and evolving systems in nature and in science and their foundational nature in our practice. But their reach goes further, into philosophical methodology. This "heuristic paradigm" (Tyson 1994) need not replace the current broader philosophical inspiration by various logical and more formalistic paradigms, which themselves have a heuristic origin in the nature of combinatorial generative systems, but it is surely an appropriate complement to them that should give us broader reach and more appropriate tools for a whole class of problems where variations may be familial rather than accidental. How to deal with intrinsically heterogeneous classes has been a problem for philosophy before and has been the origin of the idea of a family resemblance concept invoked by Wittgenstein in his *Philosophical Investigations* (1953). These problems should be expected to crop up particularly for products of evolutionary processes, where causally relevant variation is intrinsic to the mechanisms of change, whether due to selection or drift. This connection with evolutionary processes may sound uninterestingly narrow, but I include in this scope the three great systems of philosophical inquiry: *body, mind,* and *society*—all products of iterative design and selection. Donald Campbell (1974) argued that any case of fit between a system and its environment, including cognitive and cultural products, is a product of selection, and Herbert Simon basically urged the same thing in characterizing his *Sciences of the Artificial* (1969). It is a great irony that many philosophers who are refugees from engineering see that as a close call from which they have happily escaped and from which they have apparently learned nothing. And then they turn to the analysis of these three great designed systems without a clue about evolution or design and convinced that they do not need any (Wimsatt 2007; 2021a; 2021b).

So how do we proceed?

What is the common approach of analytic philosophy in evaluating a text? I was taught repeatedly, in introductory through graduate courses, to do something like the following in analyzing an argument:

1. Identify the intended conclusion.
2. Identify any premises used to argue for it.
3. Reformulate them as well-formed logical statements.

4. Organize them in the form of a deductive argument, leading to the conclusion.

5. If the conclusion does not follow, look for additional premises or alternative interpretations of the existing premises or of the meanings of the concepts in the premises that will yield a deductive argument to the conclusion.

6. If the premises are true, then accept the conclusion subject to the amended specifications and interpretations.

7. If the premises are false, indicate why, give counterexamples, and reject the conclusion.

8. If no set of true premises yielding the conclusion is found, reject the argument.

Here the whole aim is to construct a logically valid deduction of the conclusion from either the original set of premises or an extended or repaired or reinterpreted set. It is a powerful heuristic and can provide a systematic analysis and useful clarification. But it introduces a systematic bias. Usually in practice the aim of an analytic philosopher is critical: to find grounds to reject the argument and the conclusion. *This makes philosophy effectively predominantly negative and builds careers around the demolition of earlier attempts at systematic philosophy.*

But there are other productive ways in which we can approach a set of statements. The heuristics I propose for this class of problems would include the following, and as you can see, they provide plenty of analytical work for philosophers:

- Instead of looking for inexorable arguments, we look for robust tendencies and for conditions under which those tendencies are more likely to be realized.[7]
- Instead of looking for truths, we study errors and how and why they are made.
- Instead of looking for context-free inferences, we study commonly used but context-sensitive ones.
- Instead of classifying them as invalid because they are content or context specific, we should calibrate arguments to determine the conditions under which they work or are "locally valid."
- Look for other plausible assumptions of an inductive or abductive sort that may complement the existing argument.

- We may look for argument schemata, but if so, look for broad conditions where they are likely to work (like looking for the range of validity of a model) rather than trying to demonstrate their universal validity. In this way, we can espouse the use of formal methods but as a tool for appropriate problems, not as architectonic principles.
- Counterexamples become revealing sources of information about the limitations of a model or suggestions for probing its depths; in either case, they are a tool to refine the model, not an argument for trashing the system or something to be swept under the rug.
- It is thus often as important to try to refine, extend, and generalize counterexamples as it is to try to directly correct the original model. This may better illuminate the structure of failures of the original model and thus point to a deeper way to construct a new one. Similar suggestions were advanced by Thomas Kuhn (1962), and all of these preceding points were made, used, and powerfully elaborated by Lakatos in his *Proofs and Refutations* (1978). In this revolutionary work, Lakatos saw mathematical proofs as important means for the refinement of concepts in the light of counterexamples rather than establishment of the conclusion in its original form.
- For heuristics, we are looking at the adaptive structure of our cognition, or specific background assumptions, features of our social organization, or specific characteristics of the problem domain, for either strengths or weaknesses and the conditions under which these are realized. Thus, there is (or we can often extract) a reference context that contains more useful information about the method. This then recognizes methodologically the importance of context dependence.
- Rather than looking for universal theories or principles that are foundational to all other elements of a given domain, look for the conjoint application of robust principles that may be heterogeneous in application but complement each other to give a broader and richer fit to the details of the situation.
- Look for generative ways in which methodologies, empirical results, constraints, and conditions may have broad application to extend or support philosophical viewpoints, looking for the kinds of support that come from the preceding principles rather than entailments or similarly tight linkages. This should include studies of concept and meaning creation, change, and stabilization.

So the eight steps taken in analyzing an argument taught in analytic philosophy courses provide a useful and often powerful heuristic, but we should see them as that rather than an architectonic limitation of our methods. Heuristic methods permeate and constitute the vast majority of inferential tools that we have. It is time that we make a central place for them in our philosophy but also, fundamentally, in our meta-philosophy.

But if we adopt in philosophy all of these heuristic methods modeled more on scientific procedures, what is the difference between philosophy and science? This demands several distinct remarks.

First, remember that philosophy has been (and still should be) the midwife of the sciences, so the methodologies ought not be miles apart (and would not be, were philosophy not trying so hard to mark itself off as a distinct discipline). Remember that philosophy's break with psychology is less than a century old and that the new domain of science studies contains a great deal of philosophy or philosophically relevant material dealing more with social context and interaction than individual activities. It should be seen as complementing philosophy rather than replacing or destroying it.

Philosophy deals with concepts and with inference, both also the domains of psychology, but in a curious inversion, it is psychology that here has a theoretical interest in them, while philosophy as well as an abstract interest, often has a more applied focus in critiquing specific concepts and inferences! At times of significant theory change, the divide between philosophy and the sciences is harder to find because the theoretical revisions will generally involve both conceptual change and new experimental tools and methodological approaches. Thus, philosophers may increasingly require deeper studies of the science, and the scientists may be more open to philosophical input. (Biology has, within the last fifty years, been a productive source of new theoretical and conceptual directions, as this volume demonstrates. Most recently, the revolutionary expansion of the role of simulations within the last fifty years and the growing impact of big data are likely to have significant input to our methods of generating hypotheses and gathering and evaluating data and arguments [e.g., Evans and Rzhetsky 2010]).

Will philosophy disappear or be absorbed within the expanding methodologies of science as Quine predicted? I think not: philosophy should remain at one end of a continuum of methods in philosophy of science and philosophy of nature merging with natural philosophy and the sciences. And it must remain responsive to developments in those sciences. Thus, Quine's views are compromised by his attachment to methodological behaviorism. And even as epistemology, metaphysics, and meta-philosophy are affected by this expansion of biological perspectives, that still leaves ethics and value theory, logic, and history of philosophy. And the newly transformed aspects of philosophy will still remain as philosophical subjects. Philosophy is more robust and multidimensional than Quine supposed, and it becomes so in part by recognizing the role of biological practice in generating a scientific metaphysics—one with broader philosophical implications.

10. CONCLUSION: A MULTIPERSPECTIVAL REALIST METAPHYSICS

Robustness is based on a multiperspectival view of objects held in common. My metaphysical viewpoint is thus both multiperspectival and realist. This multiperspectivalism includes a multilevel mechanistic view that involves emergence and a non-eliminative articulatory reductionism. Because we are talking about practice, it is also intrinsically functional or teleological. We are intrinsically and objectively in the world, so we have no problems with the Kantian *Ding-an-Sich*. Because of entrenchment processes, it is also naturally historical and progressive. Finally, since our heuristics and means for metabolizing error are fundamentally knowledge-gaining processes, our epistemology must be fallibilistic, satisficing, and evolutionary rather than deductivist, maximizing, and foundationalist. And because we are products of evolution, this is an intrinsically naturalistic solution. So is this everything? One class of things not dealt with here is the scaffolding social, organizational, and technological interactions that midwife and extend our abilities to accomplish these activities, but that lead to an account of the nature and processes of cultural evolution, which I address elsewhere (Wimsatt 2019).

NOTES

1. As a satisficer, I would resist the other possibility—Leibniz—because of the optimality assumptions intrinsic to Leibniz's views.

2. It also removes a significant element of the supposed principled distance between rigorous scientific approaches and animal and human behavior and plans.

3. An important early inspiration for me here was von Neumann's classic and foundational essay on building reliable systems out of unreliable components (von Neumann 1956), which I read in Frank Rosenblatt's course

in 1964. A complement to this emphasizing what I have called "the metabolism of error" is Petroski's superb collection of essays (Petroski 1985).

4. This point plays a central role in my discussion of the role of ceteris paribus clauses and their ineliminability in functional assessments in Wimsatt (1972).

5. Otto Neurath famously compared the structure of science to a boat that needed to have potentially each of its timbers replaced (one at a time) while under sail. And for Campbell's "doubt/trust ratio," he vastly understates the trust required—I would put it at greater than 99 percent.

6. The notion of a heuristic was substantially developed by Herbert Simon in the late 1950s in the context of computer simulations of human behavior and decision making (his general problem solver or GPS program). Among psychologists, it was further elaborated in different directions by Tversky and Kahneman (1974), who emphasized the errors of heuristics, and in directions closer to that of Simon emphasizing the positive benefits by me and by Gigerenzer et al. (1999). Among philosophers, Thomas Nickles (2003; 2006) has also developed a systematic account of the uses of heuristics in science. My own account developed beginning in Wimsatt (1980) and is most fully elaborated in several chapters and appendices in Wimsatt (2007).

7. I first felt the need for this perspective when reading Sydney Shoemaker's *Self-Knowledge and Self-Identity* (1963) as an undergraduate. There he argued that it was a necessary condition for a language to work that people usually told the truth—something that we have seen sorely tested and increasingly validated under Donald Trump's regime. Philosophers scoffed at this new modality ("it is necessarily usually the case that") as having no acceptable semantics, but this was clearly a centrally important concept, and they should instead have scoffed at the semantic theories that could not deal with it. Again, they were misled by methodological idealizations.

REFERENCES

Avigad, J. 2020. "Reliability of Mathematical Inference." *Synthese* 198, no. 8: 1–23.

Batterman, R. W. 2021. "Multiscale Modeling in Active and Inactive Materials." In *Levels of Organization in the Biological Sciences*, edited by D. S. Brooks, J. DiFrisco, and W. C. Wimsatt, 215–32. Cambridge, Mass.: MIT Press.

Campbell, D. T. 1966. "Pattern Matching as an Essential in Distal Knowing." In *The Psychology of Egon Brunswik,* ed. K. R. Hammond, 81–106. New York: Holt, Rinehart and Winston.

Campbell, D. T. 1974. "Evolutionary Epistemology." In *The Philosophy of Karl Popper, vol. 2,* edited by P. Schilpp, 12–63. La Salle, Ill.: Open Court.

Caporael, L., J. Griesemer, and W. Wimsatt, eds. 2013. *Developing Scaffolding in Nature, Culture, and Cognition.* Cambridge, Mass.: MIT Press.

Cartwright, N. 1983. *How the Laws of Physics Lie.* Cambridge: Cambridge University Press.

Chomsky, N. 1957. *Syntactic Structures.* Berlin: Walter De Gruyter.

Corfield, David. 2010. "Understanding the Infinite I: Niceness, Robustness, and Realism." *Philosophia Mathematica* 18, no. 3: 253–75.

Eronen, M. 2015. "Robustness and Reality." *Synthese* 192, no. 12: 3961–77.

Evans, J., and A. Rzhetsky. 2010. "Machine Science." *Science* 329, no. 23: 399–400.

Feynman, R. 1968. *The Character of Physical Law.* Cambridge, Mass.: MIT Press.

Gigerenzer, G., P. Todd, and the ABC Group. 1999. *Simple Heuristics That Make Us Smart.* Oxford: Oxford University Press.

Glymour, C. 1980. *Theory and Evidence.* Princeton, N.J.: Princeton University Press.

Harper, D. 1987. *Working Knowledge: Skill and Community in a Small Shop.* Chicago: University of Chicago Press.

Herkimer, H. 1952. *Engineer's Illustrated Thesaurus.* New York: Chemical Publishing Company.

Janssen, M. 2019. "Arches and Scaffolds: Bridging Continuity and Discontinuity in Theory Change." In *Beyond the Meme: Development and Structure in Cultural Evolution,* edited by A. C. Love and W. C. Wimsatt, 95–199. Minnesota Studies in the Philosophy of Science, vol. 22. Minneapolis: University of Minnesota Press.

Kromer, Ralf. 2012. "Are We Still Babylonians? The Structure of the Foundations of Mathematics from a Wimsattian Perspective." In *Characterizing the Robustness of Science,* edited by L. Soler, E. Trizio, T. Nickles, and W. Wimsatt, 189–206. Boston Studies in the Philosophy of Science, vol. 292. New York: Springer.

Kuhn, T. 1962. *The Structure of Scientific Revolutions.* Chicago: University of Chicago Press.

Lakatos, I. 1978. *Proofs and Refutations.* Cambridge: Cambridge University Press.

Lenat, D. 1981. "The Nature of Heuristics." *Artificial Intelligence* 19: 189–249.

Levins, R. 1966. "The Strategy of Model-Building in Population Biology." *American Scientist* 54: 421–31.

Levins, R. 1968. *Evolution in Changing Environments.* Princeton, N.J.: Princeton University Press.

Nickles, T. 2003. "Evolutionary Models of Innovation and the Meno Problem." In *The International Handbook on Innovation*, edited by Larisa V. Shavinina, 54–78. Amsterdam: Elsevier Science Ltd.

Nickles, T. 2006. "Heuristic Appraisal: Context of Discovery or Justification?" In *Revisiting Discovery and Justification*, edited by Jutta Schickore and Friedrich Steinle, 159–82. New York: Springer.

Nickles, T. 2012. "Dynamic Robustness and Design in Nature and Artifact." In *Characterizing the Robustness of Science*, edited by Léna Soler et al., 329–60. Boston Studies in the Philosophy of Science, vol. 292. New York: Springer.

Odenbaugh, J. 2011. "True Lies: Realism, Robustness, and Models." *Philosophy of Science* 78, no. 50: 1177–88.

Petroski, H. 1985. *To Engineer Is Human: The Role of Error in Engineering Design*. New York: St. Martin's Press.

Shoemaker, S. 1963. *Self-Knowledge and Self-Identity*. Ithaca, N.Y.: Cornell University Press.

Simon, H. A. 1969. *The Sciences of the Artificial*. Cambridge, Mass.: MIT Press.

Simon, H. A. 1973. "On the Structure of Ill-Structured Problems." *Artificial Intelligence* 4: 181–201.

Tversky, A., and D. Kahneman. 1974. "Decision-Making under Uncertainty: Heuristics and Biases." *Science* 185: 1124–31.

Tyson, K. A. 1994. *New Foundations for Scientific Social and Behavioral Research: The Heuristic Paradigm*. New York: Allyn and Bacon.

von Neumann, J. 1956. "Probabilistic Logic and the Synthesis of Reliable Organisms from Unreliable Components," In *Automata Studies*, edited by C. E. Shannon and J. McCarthy, 43–98. Princeton, N.J.: Princeton University Press.

Wagner, A. 2005. *Robustness and Evolvability in Living Systems*. Princeton, N.J.: Princeton University Press.

Weisberg, M. 2006. "Robustness Analysis." *Philosophy of Science* 73, no. 5: 730–42.

Wimsatt, W. 1971. "Modern Science and the New Teleology: I—The Conceptual Foundations of Functional Analysis." PhD diss., Dept. of Philosophy, The University of Pittsburgh.

Wimsatt, W. 1972. "Teleology and the Logical Structure of Function Statements." *Studies in History and Philosophy of Science* 3: 1–80.

Wimsatt, W. 1974. "Complexity and Organization." In *PSA-1972* (Boston Studies in the Philosophy of Science, volume 20), edited by K. F. Schaffner and R. S. Cohen, 67–86. Dordrecht: Reidel.

Wimsatt, W. 1976a. "Reductionism, Levels of Organization and the Mind-Body Problem." In *Consciousness and the Brain*, edited by G. Globus, I. Savodnik, and G. Maxwell, 199–267. New York: Plenum.

Wimsatt, W. 1976b. "Reductive Explanation: A Functional Account." In *PSA-1974* (Boston Studies in the Philosophy of Science, volume 30), edited by A. C. Michalos, C. A. Hooker, G. Pearce, and R. S. Cohen, 671–710. Dordrecht: Reidel.

Wimsatt, W. 1980. "Reductionistic Research Strategies and Their Biases in the Units of Selection Controversy." In *Scientific Discovery, Volume 2: Case Studies*, edited by T. Nickles, 213–59. Dordrecht: Reidel.

Wimsatt, W. 1981. "Robustness, Reliability and Overdetermination." In *Scientific Inquiry and the Social Sciences*, edited by M. Brewer and B. Collin, 124–63. San Francisco: Jossey-Bass.

Wimsatt, W. 1987. "False Models as Means to Truer Theories." In *Neutral Models in Biology*, edited by M. Nitecki and A. Hoffman, 23–55. London: Oxford University Press.

Wimsatt, W. 1997. "Aggregativity: Reductive Heuristics for Finding Emergence." Philosophy of Science Association Fifteenth Biennial Meeting, 1997. *Philosophy of Science* 64, no. S4: S372–84.

Wimsatt, W. 2007. *Re-Engineering Philosophy for Limited Beings: Piecewise Approximations to Reality*. Cambridge, Mass.: Harvard University Press.

Wimsatt, W. C. 2012. "The Analytic Geometry of Genetics: The Structure, Function, and Early Evolution of Punnett Squares." *Archive for the History of the Exact Sciences* 66: 349–96. https://doi.org/10.1007/s00407-012-0096-7.

Wimsatt, W. 2013. "Scaffolding and Entrenchment, an Architecture for a Theory of Cultural Change." In *Developing Scaffolding in Evolution, Culture, and Cognition*, edited by L. Caporael, J. Griesemer, and W. Wimsatt, 177–206. Cambridge, Mass.: MIT Press.

Wimsatt, W. 2015. "Entrenchment as a Theoretical Tool in Evolutionary Developmental Biology." *Conceptual Change in Biology: The Case of Evolutionary Developmental Biology*, edited by A. C. Love. Boston Studies in Philosophy of Science. Berlin: Springer.

Wimsatt, W. 2019. "Articulating Babel: A Conceptual Geography for Cultural Evolution." *Beyond the Meme: Development and Population Structure in Cultural Evolution,* edited by A. Love and W. Wimsatt, 1–41. Minneapolis: University of Minnesota Press.

Wimsatt, W. 2021a. "Engineering Design Principles in Natural and Artificial Systems. Part I: Generative Entrenchment and Modularity." In *Philosophy and*

Engineering: Reimagining Technology and Social Progress, edited by Zachary Pirtle, David Tomblin, and Guru Madhavan, 25–52. Berlin: Springer International Press.

Wimsatt, W. 2021b. "Levels, Robustness, Emergence, and Heterogeneous Dynamics: Finding Partial Organization in Causal Thickets." In *Levels of Organization: The Architecture of the Scientific Image*, edited by Daniel S. Brooks, James DiFrisco, and William C. Wimsatt, 21–38. Cambridge, Mass.: MIT Press.

Wimsatt, W., and J. Griesemer. 2007. "Reproducing Entrenchments to Scaffold Culture: The Central Role of Development in Cultural Evolution." In *Integrating Evolution and Development: From Theory to Practice*, edited by R. Sansome and R. Brandon, 228–323. Cambridge, Mass.: MIT Press.

Wittgenstein, L. 1953. *Philosophical Investigations.* Oxford: Blackwell.

2

HOW TO INFER METAPHYSICS FROM SCIENTIFIC PRACTICE AS A BIOLOGIST MIGHT

WILLIAM C. BAUSMAN

1. FROM BIOLOGICAL PRACTICE TO SCIENTIFIC METAPHYSICS?

It is time to address the epistemological problem of how we are to make inferences, from premises involving our analyses of scientific practices to conclusions of metaphysical claims about the world. Our need for a new model of inference stems from conjoining practice-based epistemology of science with scientific metaphysics. On the Quinean picture of scientific metaphysics, we can glean the ontology of the world by analyzing the objects quantified over in the statements of our best scientific theories (Quine 1948). Representing inferences to metaphysical claims from scientific theories is logically simple because both are propositional and truth-apt. The problem for a practice-based scientific metaphysics is how to model inferring from a practice, including tools and methods that are not truth-apt, to a metaphysical claim.

Ian Hacking advocated analyzing how science succeeds in terms of intervening and not representing. His slogan "If you can spray them, then they are real" (Hacking 1983, 23) is an example of what an inference pattern from a practice to a metaphysical claim could look like. However, the form of and support for Hacking's simple inference rule remain mysterious (Miller 2016).

I am interested in arguments of the following form: 'The practice goes so-and-so because the world is such-and-such' or 'We know the world is such-and-such because the practice investigating it goes so-and-so.' The problem facing the epistemology of scientific metaphysics here is that we need then a model of inference based on a relation that holds between a practice and the world. Ken Waters recently made a bold metaphysical argument about the structure of the world based on how biologists practice genetics: "My metaphysical claim is that scientific practices in genetics and allied sciences take this form because they are adapted to a reality that has no overall structure" (Waters 2017, 99). I analyze Waters's reasoning at length in this chapter. In addition to being a clear example of making the kind of argument I am interested in, Waters inadvertently suggests an exciting and promising solution to our problem. What if successful scientific practices were *adapted* to features of the world? If features of a scientific practice were adapted to the world, then it would perhaps be possible to learn about the world by investigating the practice. This proposal has promise because, rather than developing a novel model of inference for metaphysics based on adaptation, we can draw on analogous inferences used by biologists to solve analogous problems.

The analogy I suggest we take seriously then is: Successful scientific practices are adapted to features of the world for purposes as organisms are adapted to features of their environment for functions. Biologists use traits of organisms as proxies for environmental conditions. My metaphysical methodology is therefore to pursue the use of traits of scientific practices as proxies for metaphysical features of the world. I devote this chapter to developing the beginnings of what this research program looks like and how it functions. Biological practice should inform both the methodology and content for doing scientific metaphysics. We should use this biologically inspired methodology to explore the metaphysics of any parts and aspects of the world we can get a grip on through analyzing the practices of all the sciences.

Paying attention to the biological practice for methodological advice promises to provide guidance for dealing with the formidable, basic problems facing a scientific metaphysician such as Waters. If we take the talk of adaptation seriously, we should also accept that, if the world were structured differently, the practice of genetics would be different. How can the metaphysician give evidence for the ability of a scientific practice to track the structure of the world? If we accept that scientists can shape their investigative practices to fit their environment, why think that the reason a practice is the way it is is because of the metaphysics of the world and not because of the history or sociology of the discipline? Perhaps genetics uses a particular gene concept not because the structure of heredity lacks structure but because of the gene concepts that came before it and because of how new biologists are educated and trained. There are many alternative hypotheses as to why a given scientific practice is the way it is, and the scientific metaphysician must show that the adaptation hypothesis is better supported. In each of these problems and more, biologists face an analogous problem and have developed accepted ways for answering it.

I develop my proposal as follows. In section 2, I present how paleoecologists infer to past environmental conditions from fossils of organisms that lived in them. I use the case of inferring from the shape of teeth to past humidity to draw lessons about inference patterns and support to construct the proxy by adaptation framework. In section 3, I import the proxy by adaptation framework to practice-based scientific metaphysics. I use the case of Ken Waters inferring from geneticists' use of the gene concept to the structure of heredity, development, and evolutionary change to explore how we can apply the framework. In section 4, I comment on several decisions I made in developing the proxy by adaptation framework. In section 5, I discuss the relationship between science and metaphysics on my proposal for a biologically inspired, practice-based scientific metaphysics.

2. FROM ORGANISM TO ENVIRONMENT

How do scientists learn about inaccessible environments? When they cannot directly measure environmental variables, scientists construct and measure environmental proxies. Proxy variables are proxies for other variables based on some understood physical, chemical, or biological processes. Geologists take ice core samples to measure the relative ratios of gases trapped in bubbles. They use gases as a proxy for atmospheric temperature based on physical and chemical processes of isotopes. Geologists also use the levels of dust in ice as a proxy for dryness of the environment because wind blows the dust produced by erosion around more in dry areas (Bender, Sowers, and Brook 1997). Biologists use tree rings as proxies for past temperature and dryness based on ecological and physiological conditions favoring and disfavoring growth. Paleoecologists use pollen counts in soil as proxies to reconstruct past climate based on ecological dispersal and competition. They must identify, calibrate, and evaluate these proxies in an iterative process of continual refinement.

Sailors use seabirds as a proxy for land when they are at sea based on experience and ecology. Alfred Russel Wallace used the long nectary spur of an orchid from Madagascar as a proxy to infer the existence of a pollinator moth with a long proboscis that sucks it. Wallace (1867) made his inference on the basis of adaptation via natural selection.[1]

In order to articulate the proxy by adaptation framework of reasoning used by biologists, I examine the case of mammalian teeth as proxies for past humidity. Biologists know an extraordinary amount about mammalian teeth.[2] Teeth are the most durable and well-preserved parts of animals, and they are one of the most morphologically diverse. Because of their functional importance, they are highly specialized and distinct across even closely related taxa and strongly subject to adaptation by natural selection.[3]

Mikael Fortelius (1985), professor of evolutionary paleontology at the University of Helsinki, is an expert on Cenozoic (last 66 million years) ungulate (hoofed animals from horses to hippos to deer) teeth. Fortelius leads a multifaceted research program using teeth and other traits to learn about past climate change. I focus here only on the group's use of one trait of ungulate teeth—their hypsodonty, or height (Figure 2.1). The following is my reconstruction of their reasoning.

To begin, a team digs up teeth and records their location. They date the teeth using the geological context they find it in and other proxies for age, itself a complex procedure. At first, Fortelius and team look for unworn teeth, ideally second upper molars, of known species. They measure features of the teeth including the molar crown width and crown height and use these to classify the species into one of three classes of increasing height: brachydont, mesodont, hypsodont. Species with high molars are called hypsodont and show more hypsodonty than species with lower molars. They then assume that all members of a given species have teeth in the same height class. This allows them to track changes in tooth size using other databases of species, not necessarily based on tooth specimens. They also track the variation in found tooth specimens to observe evolution change and to limit their uniformity assumption.[4]

With this information in a database, biologists track changes in ungulate teeth shape over space and time. They have global data for the Neogene period from 24 to 2.5 million years ago. They then generalize from the data.[5] One particular generalization is that mean hypsodonty increased in herbivores in the late Miocene (10.5–5 mya) in Europe (Fortelius et al. 2006).

Next, using a form of abductive inference, they hypothesize that increased hypsodonty in herbivores is an adaptation to eating plants in an environment with decreased water[6] (Fortelius et al. 2002; 2006). Using this hypothesis, they infer to the condition: Water in the environment decreased in the late Miocene in Europe. Their full inference runs as follows:

Figure 2.1. Hypsodonty. These partially fossilized horse molars are very hypsodont. Photographed by Derby Museums Trust, Rachel Atherton, courtesy of The Portable Antiquities Scheme/ The Trustees of the British Museum.

TEETH-WATER INFERENCE

1. Hypsodonty increased in herbivores in the late Miocene in Europe.
2. Increased mean hypsodonty in herbivores is an adaptation to eating plants in an environment with decreased water.
3. Therefore, water in the environment decreased in the late Miocene in Europe.

The inference pattern used here is strong. Its basic form is:

Adaptation Inference Schema
Descriptive claim: Organism O has trait T.
Adaptation hypothesis: Trait T in organism O is an adaptation to feature
    F of environment E for function N.
Conclusion: Therefore, environment E has feature F.

A trait is adapted to an environment if it arose in that environment by natural selection. This is why adaptive traits are suited to being proxies, the only problem being that organisms can continue to carry traits after their environment changes. This issue shows that both the first premise and conclusion in an adaptation inference such as the teeth-water inference need to be carefully dated. The cost of a historical definition of adaptation is increased evidence needed to justify the adaptation hypothesis used as the second premise in an adaptation inference.[7]

Putting aside the significant empirical justification needed for the descriptive claim, the main scientific work in learning about the past using teeth is justifying the adaptation hypothesis that they abduced. This abduction does not itself give evidence, but rather guides their future research. To support the adaptation hypothesis and to make it exportable to other scientific contexts, Fortelius's group tries to establish the following four claims:[8]

A. Establish that a **correlation** holds between mean hypsodonty and amount of water in other places and times.
B. Support that increased hypsodonty is **functionally adaptive** to eating plants when humidity decreases.
C. Support that increased hypsodonty was **selected** for in herbivores eating plants in Western Europe in the late Miocene given decreased humidity.
D. Establish the **robustness** of their conclusion using other proxies.

In the remainder of this section, I explain how they go about investigating these four claims. In addition to presenting the form of a strong proxy inference, I present enough detail about Fortelius's research program to function as an exemplar to be followed when we move to scientific metaphysics. Together, this inference schema and these four claims form the basis of the proxy by adaptation framework that will be our guide to making good in-ferences in scientific metaphysics, as I show in section 3.

2.1 Correlation

The basis for using a trait as a proxy is finding a correlation between the trait and the environmental condition it is a proxy for. However, if the envi-ronmental condition could be measured directly, no proxy would be needed. The solution to this puzzle is to correlate the trait with the environmental condition in a different context where it can be measured.[9]

Fortelius's group finds a negative correlation between mean hypsodonty and humidity in contemporary data from direct measurements of the hu-midity (Eronen, Puolamäki, et al. 2010a; Damuth et al. 2002). Statistical cor-relation further allows them to calibrate a proxy. Calibration is the process by which a functional relationship is established between the changes in the proxy variable and the unmeasured variable (Eronen, Puolamäki, et al. 2010b; Eronen, Polly, et al. 2010; Polly et al. 2011).

Correlation alone is a weak justification for a proxy. The justification for exporting the correlation found in one context to another context comes from understanding the conditions under which and processes by which the correlation holds.

2.2 Functional Adaptation

A trait is functionally adaptive to a feature of an environment if and only if the trait is good for that function in the environment. Functionally adaptive traits are sometimes called "aptive" traits because they do not rely on being produced by natural selection but only the fit to the environment (Gould and Vrba 1982).

Fortelius argues that many major structural differences of teeth between groups, including hypsodonty, are related to specific differences in their functional demands (Fortelius 1985, 64). Hypsodonty is aptive for mam-mals eating grasses because grasses contain phytoliths, internal silica par-ticles, and can also be covered in grit, both of which will grind down teeth quicker than without it. Their teeth do not grow continually or regenerate.

This can be directly observed in experiments using either just teeth or ungulates with varying degrees of phytoliths and grit in the food. This provides both a basic test of the aptation claim and a measure of the rates of wear.[10]

2.3 Selection

While increased hypsodonty is aptive for eating plants in arid environments, it could be either an adaptation or an exaptation (former constraint built into a functional system). A trait is an adaptation for a function if the trait is aptive for a function and was selected for that function because of being aptive. A trait is an exaptation if it is aptive for a function and was not selected for that function but arose in a different environment, as a developmental constraint, or as linked to some other adaptive trait (Gould and Vrba 1982). Fortelius argues that teeth are a complex construction of adaptations and exaptations (Fortelius 1985).

That a trait is aptive is not enough to support its use as a proxy. Any given trait of an organism will have some functionality in any given environment. And some novel environments would be great places for it to live. Consider invasive animals prior to invasion. The problem facing paleobiologists is that, while ecologists can tell us in which environments a species might do well, such a prediction alone does not inform where and when it lives. When a trait is an adaptation to a function in an environment, you know that the species with the trait lived in that environment.

Fortelius's group provides several lines of argument for why increased hypsodonty is an adaptation to decreased water. They invoke two kinds of selection: evolutionary and ecological. And they offer two kinds of support for each kind of selection: how possible and convergence arguments.

2.3a How Possible

How possible arguments for adaptation via natural selection explain how an aptation could come about via a series of small adaptations. Darwin offered these to explain, for example, the evolution of caste structure in ant colonies in the *Origin,* and they remain important.

Decreased water in the environment would lead to increased hypsodonty in populations of herbivores via natural selection by the following steps:

1. Decreased water in the environment causes more fibrous plants, more open landscapes, and plants with more grit on them (Janis and Fortelius 1988).
2. Herbivores need to eat plants and need teeth to eat plants.
3. Herbivore teeth grind down faster when eating more fibrous plants with more grit, eventually below a functional level (Janis and Fortelius 1988).

4. Herbivore teeth cannot quickly become more durable (Janis and Fortelius 1988; Eronen, Puolamäki, et al. 2010a, 218).

5. Therefore, if water decreased in the environment and herbivore teeth didn't grow more hypsodont, then herbivores would not live as long.

6. Herbivore teeth naturally vary in hypsodonty, and this variation is heritable.

7. Therefore, increased hypsodonty will be selected for when water decreased.

8. Therefore, increased hypsodonty in herbivores is an evolutionary adaptation to decreased water in the environment.

In this way, over evolutionary time, new variation including teeth with greater hypsodonty will continually arise. The mean hypsodonty in a population will increase as water in the environment decreases and hypsodont teeth are selected for.

Fortelius's group actually argues that increased hypsodonty is primarily *ecologically adaptive* to decreased water. A trait is ecologically adaptive when it is ecologically aptive and when it was caused by ecological selection, also called interspecific competition and competitive exclusion (Vellend 2010). This means that if a novel species of herbivore immigrates into a community in which it has a relatively greater mean hypsodonty, the species will tend to outcompete the other herbivores for limited resources over ecological time (Fortelius et al. 2006).

The reason that ecological adaptation is more important than evolutionary adaptation is because the increases in hypsodonty are driven by the commonest species across Eurasia (Jernvall and Fortelius 2002). The species turnover in communities during mean hypsodonty increases suggests that it is the ability of hypsodont species to migrate and outcompete other species that drove the increase in mean hypsodonty.

One important difference between thinking evolutionarily and ecologically is the time scale, and this has implications for the types of change that are possible. Another difference is a shift from the population to the community as the unit of analysis.[11] The changes in phenotype in one population that happen evolutionarily cannot happen ecologically to a community. But phenotypes can change in one area via ecological processes because the particular species in the area are not fixed. Also, the source of variation is different. Mutation is the main source of evolutionary variation, while dispersal is the main source of ecological diversity.

Analogously, decreased water in the environment would lead to increased hypsodonty in populations of herbivores via ecological selection by the following steps:

1. Decreased water means both more fibrous plants and plants with more grit on them.
2. Herbivores need to eat plants and need teeth to eat plants.
3. Herbivore teeth will grind down faster when eating more fibrous plants with more grit.
4. Herbivores cannot easily migrate to regions with more water.[12]
5. Therefore, if water decreases in the environment and the populations didn't migrate to wetter pastures, then herbivores' lifespans will shorten and their reproductive success decrease.[13]
6. There is dispersal of herbivores with variation in hypsodonty and location is maintained by offspring.
7. Therefore, herbivores with greater hypsodonty will outcompete herbivores will lesser hypsodonty.
8. Therefore, increased hypsodonty in herbivores is an ecological adaptation to decreased water in the environment.

In this way, over ecological time dispersal will introduce herbivores with large ranges and with greater hypsodonty to a community. The mean hypsodonty in the community will then increase as water decreases and as these hypsodont migrants outcompete the locals.[14]

2.3b Evolutionary Convergence

One the strongest signs that a trait is an adaptation is finding the convergence of a trait from independent lineages in similar environments and for the same function. For example, that dolphins and sharks have the same streamlined body shape is a strong sign that this shape is an adaptation to hunting in the water.[15] But correlations across ecological communities between hypsodonty and water are also a kind of convergence argument for ecological adaptation.[16]

Using correlation data, the biologists can show that hypsodonty in herbivores evolved independently in North and South America, Europe, Asia, and Africa. This is unlikely to have happened due to chance or developmental constraints. Therefore, it is probably an adaptation to foraging with decreasing water.

Together, showing a correlation, functional adaptation, and selection via both how possibly arguments and convergence arguments produces a strong argument that a trait is an adaptation. This justifies the adaptation hypothesis in the inference from trait to environment. The final step in justifying the conclusion about the environment comes from checking the conclusion for robustness.

2.4 Robustness

The conclusions drawn from one proxy variable should be checked with those drawn from another proxy variable that has already been confirmed and calibrated. Trends about temperature drawn from ice core samples and tree rings for the same place and time should agree. This is a form of triangulation or robustness of the conclusion from particular premises (Campbell and Fiske 1959; Wimsatt 1994). This is an important step in establishing that a variable can be a proxy variable for another variable and for calibrating the functional relationship between the variables.[17] The stated primary function of developing hypsodonty as a paleoprecipitation proxy is its use in vegetation and climate models (Fortelius et al. 2002). The group has confirmed its broad agreement with other climate proxies (Eronen, Puolamäki, et al. 2010b; Liu et al. 2012; Žliobaitė et al. 2016).

Now that we see how biologists use adaptation to make inferences from observed organism traits to unknown environments, we can begin to apply these lessons to doing scientific metaphysics.

3. FROM PRACTICE TO METAPHYSICS

I began this chapter with a research proposal based on an analogy between the adaptation of organisms to the environment they live in and the adaptation of scientific practices to the world they investigate. We learned from analyzing paleobiological practice that there is a straightforward way to make inferences from traits of organisms to their environments using adaptation. And I presented four claims A–D that, when established, justify the adaptation inference: showing correlation, functional adaptation, selection, and robustness. My task in this section is to apply the analogy and its lessons to doing metaphysics: we should seek to establish metaphysical proxies in scientific practice on the basis of adaptation.

To show how to do this, in this section I fit as well as possible a case of scientific metaphysics from biological practice into the proxy by adaptation framework. My case study is Ken Waters's (2017) argument for the no general structure thesis.

Waters (1994; 2004) has long investigated the practices of classical and molecular genetics from an epistemological perspective. He is interested in why genetics succeeds and proposes that genetics succeeds not because of a core theory but because of its investigative practices (Waters 2019). But more recently he went beyond an epistemological understanding of scientific success and entered the realm of metaphysics. In "No General Structure," Waters (2017, 99) argues that the success of genetics is explained in part by the no general structure thesis: "Reality has lots of structure, but no overall structure."

What exactly metaphysicians mean when they use "structure" is notoriously slippery. To gain enough traction to make progress fitting Waters's reasoning for his conclusion into the proxy by adaptation framework, here is one way of understanding what Waters means by "structure."[18] We can distinguish between two kinds of structure: horizontal and vertical structure. Consider these concepts using an object that has one version of perfectly general structure—a Sierpinski triangle (Figure 2.2). A Sierpinski triangle is a fractal that is self-similar: if we divide the largest triangle into four triangles of equal area, each of the subtriangles has the same structure as the whole. And this continues at smaller and smaller scales because the pattern is repeated downward infinitely.

Horizontal structure is the structure at one spatial and temporal scale. About horizontal structure, we ask, Are their basic units at a scale? In a Sierpinski triangle, the answer is clearly yes at every scale. Vertical structure is the structure across scales. About vertical structure, we ask, Does the structure of one scale reduce to the structure of another scale? Again, in a Sierpinski triangle, the answer is clearly yes between any two scales. The question about structure for Waters is whether the parts of the world investigated by geneticists are structurally similar to this idealized fractal structure. While similarity is a quantitative relation, for simplicity I discuss it only as a qualitative relation of having or lacking horizontal or vertical structure. *General* structure is a claim about a degree of structure and can be applied to both horizontal and vertical structures.

I reconstruct Waters's road to the no general structure thesis as follows:

Step 1: Adopt the practice-centered view of science.
Step 2: Analyze the practice of genetics from the practice-centered view.

Figure 2.2. Sierpinski triangle.

Step 3: Make inference 1.

Step 4: Conclude first that the domain investigated by the practice of genetics lacks a horizontal structure.

Step 5. Make inference 2.

Step 6: Conclude second that the world lacks vertical structure.

Inference 1 to the first conclusion is where Waters infers from biological practice to scientific metaphysics. Inference 2 on my reconstruction is an inference from a metaphysical claim about one domain to a metaphysical claim about the whole domain. Only steps 1–4 centering on inference 1 and its sources of justification concern us here.

In step 1, Waters begins his reasoning from the practice-centered view of science. One purpose of this volume is to better explore and understand the practice-centered view of science. But for Waters, this view is about shifting the analysis of science from foregrounding theories as things to foregrounding practices as activities. A useful slogan for the practice-centered view is not theories but theorizing, not models but modeling, not experiments but experimenting.[19]

In step 2, Waters draws on his analyses of the practices of both classical and molecular genetics to make his argument. However, the basic form of his argument for the no general structure thesis can be seen from geneticists' use of the molecular gene concept. Geneticists use the molecular gene concept to investigate the domains of heredity, development, and evolutionary change. I use "heredity" as a shorthand for all of these domains as Waters's claims are the same for each.

Waters argues that geneticists use the molecular gene concept when they want to be precise: "A gene g for linear sequence l in product p synthesized in cellular context c is a potentially replicating nucleotide sequence, n, usually contained in DNA, that determines the linear sequence l in product p at some stage of DNA expression" (Waters 2017, 95). The practical consequence of using the molecular gene concept to understand heredity is that it parses different linear sequences into the genes for different cellular products. Therefore, there is no set of genes from which everything is constructed. Waters describes this using his epistemological concept of a fundamental unit: "What does it mean to say the gene is the *fundamental unit* of heredity? Presumably it means that if you could identify every gene and every difference in every gene, and if you could trace the transmission of each gene and each gene difference from one generation to the next, then you would have a comprehensive basis for understanding everything about heredity" (Waters 2017, 92). Waters argues that the molecular gene concept does not function as a fundamental unit of heredity because the molecular gene concept does not uniquely parse linear sequences into cellular products.

The conclusion of Waters's analysis of the practice of genetics from the practice-centered view of science is the practice of genetics' use of the molecular gene concept does not provide a single correct parsing of DNA into genes for understanding the domain of heredity.

In step 3, Waters draws on his analysis of practice to infer his first metaphysical conclusion. His conclusion is the domain of heredity lacks a horizontal structure. But we are interested in his argument for this claim. Waters twice invokes adaptation to explain his conclusion: "My metaphysical claim is that scientific practices in genetics and allied sciences take this form because they are adapted to a reality that has no overall structure. The reality has lots of structure, but no overall structure. Practice has been adapted to work in the reality of the world that biologists are engaging" (Waters 2017, 98). While Waters uses the language of "adaptation," he does not elaborate on what he means by it or what work it is doing in his argument. Such loose talk of adaptation and fit is common among philosophers when discussing the relationships between things and the world. Drawing on the analysis of paleobiological practice though, we can investigate how the reasoning goes if we read "adaptation" as analogous to biological adaptation. Waters's adaptation hypothesis about genetics and structure then is the practice of genetics' use of the molecular gene concept is an adaptation to manipulation in a domain that lacks a horizontal structure.

Waters's inference 1 then runs as follows:

GENE–STRUCTURE INFERENCE
1. The practice of geneticists' use of the molecular gene concept does not provide a single correct parsing of DNA into genes for investigating the domain of heredity.
2. The practice of genetics' use of the molecular gene concept is an adaptation to manipulation in a domain that lacks a horizontal structure.
3. Therefore, the domain of heredity lacks a horizontal structure.

For our purposes here, we can take for granted that the descriptive claim in the first premise is justified and focus on the justification for the adaptation hypothesis in the second premise. Waters himself argues for premise 1 extensively but does not provide any direct evidence for premise 2. Applying the proxy by adaptation framework we developed from paleobiology, four claims need to be evidenced for this inference to run smoothly and the conclusion to be exportable. They are:

A. Establish that a **correlation** holds between parsing the domain and horizontal structure in other, independent practices and domains.
B. Support that the molecular gene concept is **functionally adaptive** to manipulation in a domain without a horizontal structure.
C. Support that the molecular gene concept was **selected** for in genetics for the purpose of manipulation given no horizontal structure.
D. Establish the **robustness** of the lack of a horizontal structure in heredity, development, or evolutionary change.

In the remainder of this section, I discuss the prospects for showing these four claims. Waters's argument is my special target, but I also speak more generally about meeting these standards for any metaphysical argument.

3.1 Correlation

We should establish a correlation between the trait of the practice being used as the proxy and the metaphysical feature of the world it is a proxy for. However, if this could be done directly for the metaphysical feature of the world, then no proxy would be needed as we would already have what we want. When we need a proxy, we seek a correlation between the same kinds of things as in the case under investigation. Then an inference can be made on the assumption that the same relationship holds between scientific practice and metaphysics.

In the metaphysics case at hand, we need to correlate domain parsing and horizontal structure in cases where we have independent access to both. We need not find a case of *direct* access to the metaphysical horizontal structure, which we do not have, but rather *independent* access. The sense of independence needed is independence from the particular epistemic analysis of the practice being analyzed to draw metaphysical conclusions. Independence from other epistemic analyses of science itself is not required. The case also needs to be relevantly similar in background conditions to make it reasonable to generalize to the metaphysics case.

Paleobiologists correlate teeth of ungulates and humidity in the present because they can directly observe the present and because they think that ungulate teeth and water still interact in the same way. The accessibility of the past is limited, but because the present resembles the past, they reason back using actualism and local empirical support for it.

I see two complementary ways of getting a grip on the metaphysical structure of the world, each analogous to actualism. First, we can begin with our grip on the everyday structures. Arguably our concept of metaphysical structure is based on concepts of structure that we are more familiar with. If a correlation can be established between ways of parsing domains for functions and horizontal structure of a domain in accessible cases, then it can be carefully extended beyond the accessibility horizon using background knowledge, some basic assumptions analogous to actualism in historical sciences, and the process of iteration.

Structure is found in any spatially or temporally extended system. An extended structure has horizontal structure to the degree that any one part is similar to another part considered at a scale. In explaining what he means by general structure, Waters gives the example of cities. Calgary has more horizontal structure than Arles does because Calgary was constructed on a grid pattern while Arles grew historically. In Calgary, if you know how to navigate one quadrant, you have a good idea how to navigate any other quadrant because the streets and avenues maintain their cardinal orientation for the most part, while in Arles no neighborhood is a good guide to getting around any other. Applying Waters's categories, we can say that in Calgary, the streets and avenues behave like a fundamental unit. Streets and avenues also parse Calgary uniquely in that no matter if you want to navigate by foot, bicycle, or car, you can use the same directions. This is not true if you consider light rail or freight transportation.

An example of temporal horizontal structure comes from blues and jazz songs with repeating chord progressions. In a simple twelve-bar blues song, you cycle through the I I I I, IV IV I I, V IV I I chord progression a few times, and that is it. These twelve bars are the fundamental unit of the song, and they will parse the domain uniquely. More complex songs will change the chord progression and introduce additional parts of the song—for example, an ABA song structure. But most songs still have high horizontal structure and require this so that both the players and the listeners can follow along. Songs with low horizontal structure are often composed as such on purpose to break out of conventions.

The second way to get a grip on metaphysical structure is by abstracting from our construction of simple mathematical and formal objects. I used the Sierpinski triangle previously to introduce the concept of structure. It is a perfectly self-similar fractal, the same at a scale and across scales. It can be constructed in any number of ways—for example, as the limit of smaller and smaller triangles or squares. Either a triangle or a square then would work as a fundamental unit and would parse the domain uniquely. Most fractals, however, are not self-similar, and there are various ways of measuring this. Their unifying feature is their noninteger dimensionality and roughness across scales. Fractals are useful for measuring the length of coastlines, for example.

To summarize, we face a version of Meno's Paradox of Inquiry: if we have access to metaphysics, correlation is unnecessary, while if we do not have access to metaphysics, correlation is impossible. Therefore, correlation is either unnecessary or impossible. We can analyze this paradox into two problems. First is the access problem: How do we gain independent access to the metaphysics of the world? Second is the meaning problem: How do we understand what our metaphysical hypotheses mean? My solution to both is iteration from our everyday experience and formal constructions. I propose that we continue to use the same iteration technique that we use when designing scientific instruments as elaborated by Hasok Chang (2004; 2007). This can then be supplemented with other forms of inquiry throughout the iteration process.

3.2 Functional Adaptation

Waters needs to support that the molecular gene concept is functionally adaptive to manipulation in a domain without a horizontal structure. More carefully, he needs to show that gene concepts that do not parse the domain uniquely work better than those that do to support manipulation in a domain lacking a horizontal structure compared with having a horizontal structure. Waters gives no reasons to support functional adaptation, and I am aware of no other metaphysicians who have done so for their cases either. It is left as a suggestive hypothesis without direct evidence for it given.

Within the proxy by adaptation framework, the question is then how to measure the fitness of traits of scientific practices in a domain/environment. The main difficulty for showing this is a corollary of the access problem: the domain that a scientific practice investigates cannot be varied in a controlled experiment or found to vary naturally. One prospect for investigating the fitness of scientific practices in different environments is computer modeling and simulation, provided we can construct a fitness measure. In a computer, we can construct a world with any structural architecture desired. Evolutionary models of fitness can themselves be adapted to understand the fitness of scientific practices. This could inform us that certain traits are more aptive in certain environments than others.

3.3 Selection

Assuming functional adaptation, Waters needs to show that geneticists came to not parse heredity into unique fundamental units because parsing this way had higher relative fitness for manipulation. This can be substantiated using both how possible and convergence forms of analysis.

3.3a How Possible

A story needs to be given that explains how the molecular gene concept could have possibly come about due to selection operating on variation in the practice of genetics. We saw that paleobiologists distinguish evolutionary and ecological forms of selection and adaptation. However, the differences in thinking evolutionarily versus ecologically about scientific change have yet to be developed. Again, the relevant differences between evolution and ecology are the units of analysis (population versus community), time scale (longer versus shorter), and the source of variation (mutation versus dispersal). I think that in general historians and philosophers are too quick to think evolutionarily rather than ecologically about scientific change, but I will not argue this further here. In what follows, I speak neutrally between an evolutionary and ecological kind of change.

The actual change of the gene concept is known to historians. This can be used for staging its development. However, the explanation also must explain why the practice of genetics would have been different if the structure of the domain of heredity was different. Knowing the past is not sufficient to answer this question.

Here is an example of the kind of how possible argument required to establish selection, with the component claims identified:

1. The lack of horizontal structure in the domain of heredity means the lack of a fixed set of joints.
   (Definition)
2. Geneticists need to interact with the structure of their domain via their conceptual practice.
   (Statement of interaction)
3. Gene concepts that are rigid are not good for describing a structure that lacks fixed joints.
   (Statement of ill fit)
4. The gene concept cannot grow less unified (meaning cannot just completely dissolve into purely local concepts) for practical reasons of education and communication.
   (Statement of nonviability of alternative solutions)
5. Therefore, if their domain lacks horizontal structure and if genetical conceptual practices did not grow more flexible, then geneticists could not make increasingly successful investigations.
   (Summary of pressure)
6. There was variation in the flexibility of the classical gene concept.
   (Statement of variation)

7. Therefore, increased flexibility of the gene concept will be selected for when its domain lacks horizontal structure.
(Summary of selection)

8. Therefore, the practice of genetics' use of the molecular gene concept is an adaptation to manipulation in a domain that lacks a horizontal structure.
(Sought adaptation hypothesis)

In this way, over time, new variation including a more flexible gene concept could arise. A gene concept that does not uniquely parse its domain will come about in an environment lacking a horizontal structure as flexibility in the gene concept is selected for.

3.3b Convergence

In contrast to how possible adaptation explanations, convergence gives "that actual" adaptation explanations. We can analyze an adaptation hypothesis into two parts:

(i) X is an adaptation.
(ii) X is an adaptation to Y for function N.

Convergence by itself is a sign of (i) only. Knowing that two traits of distantly related species converged in their evolutionary trajectories is evidence only that the two traits are adaptations, not what the traits are adapted to or for what function.

The special problem for showing convergence concerns which comparison classes to use to show convergence. With organisms, convergence is only a sign of adaptation when the organisms are not closely related. If the organisms are closely related, then their traits are more likely homologs than analogs. Shared traits are either homologs or analogs depending on whether the trait is shared in virtue of common descent (homologs) or not (analogs).

The special question is then what notion of common descent in scientific practices is needed to understand trait convergence as a sign of adaptation? If two labs studying heredity with the classical gene concept were to both come to use the molecular gene concept, would this count as convergence? It would be more significant if the molecular gene concept were found to be used in lineages of practices with more independence. However, there is nowhere near as much independence in science compared to life.

3.4 Robustness

The fourth claim needed when establishing a proxy is its agreement with other established proxies of the same environmental feature. Teeth as a proxy for water are checked with tree rings as a proxy for water. Domain parsing as a proxy for horizontal structure should then be checked with other proxies for horizontal structure in the same domains.

The main difficulty with using robustness analysis is that it requires other proxies. The simplest case would be to check a new proxy against an already-established proxy. But additional proxies need not already be fully established, for it might be that two proxies for the same metaphysical feature are investigated concurrently.

Another potential use of robustness is to admit other forms of philosophical analysis to compare with metaphysical proxies, such as linguistic and a priori analysis. This suggestion will be controversial to some depending on your metametaphysical commitments, but there is nothing inconsistent about using multiple approaches. A strong reason for doing scientific metaphysics from biological practice is that it is an unused source for doing metaphysics, but there is no reason that used sources need to be rejected when new sources are found. There need not be only one kind of metaphysical methodology. What is then needed here is a method for combining different sources of justification and evidence.

Finally, with environmental proxies in science, robustness is used to both establish and calibrate a proxy. Calibration is used to determine the functional relationship between a proxy variable and its environmental variable. The idea of calibrating a proxy should be explored for metaphysics. Some metaphysical proxies will certainly be at least qualitative. In my analysis of Waters's no general structure thesis, I have simplified the idea of structure to be a total presence or absence question, but most forms of structure admit of more and less and even degrees.

4. COMMENTARY

In this section, I comment on choices I have made regarding inference, success, and adaptation in developing this proposal.

4.1 Inference

The first decision I made regarding the inference from practice to metaphysics was to use a proxy model of reasoning and the proxy by adaptation framework. My reasons for this are based on the ability it gives to draw from scientific practice in making inferences to environments we do not know much about. But it is helpful to contrast this approach with another to appreciate it.

The main alternative approach that I considered is the exportation model of inference. Understood this way, realist metaphysics is a matter of exporting the buried metaphysical claims outside of the scientific practices. On this account, the basic metaphysical inference goes:

EXPORTATION INFERENCE SCHEMA
1. Scientific practice P behaves as if situation S holds. / P accepts S.
2. P is successful.
3. Therefore, S.

S here is a proposition such as "The world lacks a general structure." and ". . . is successful" as shorthand for whatever property the practice has that makes it worthy of exportation.

One example of doing metaphysics using an exportation model is Carl Gillet's (forthcoming). He distinguishes between *internal ontology* and *ultimate ontology*. Internal ontology is a descriptive enterprise of what goes on in science, while ultimate ontology is a normative enterprise about what there really is and which requires substantial further justification.[20] The justification required, in my view, can be framed in terms of making good exportation inferences. Other philosophers have held similar views; for example, Kuhn (1962) identifies metaphysical assumptions as an important part of paradigms.

My issue with the exportation model of inference is that it hides the issue of where we get the "S" from. I do not see identifying metaphysical assumptions or descriptive ontology as a straightforward exercise in the description of scientific practice. In short, I do not think we should see metaphysical assumptions as only brute assumptions of a practice, but rather as linked to other methodological practices. An important part of the metaphysical project for me is understanding where the metaphysical hypotheses even come from, how we link them to practice, and how we understand what they mean for investigation. And therefore, in order to make the full line of reasoning plain, we should begin the reasoning from descriptions of practice alone. I do not believe that the exportation model of inference necessarily relies on metaphysical assumptions being easy to read off scientific practice, and I suspect that everything I propose in this chapter using proxy-based inference could be more or less adapted to fit the exportation model. Representing the reasoning with the proxy model is plainer, more complete, and closer in structure to inferences used by biologists.

The second point concerns how to arrive at the adaptation hypothesis used in the adaptation inference. As I hope is now clear, biologists do not use inference to the best explanation to support adaptation hypotheses. To use inference to the best explanation is to take the explanatory power of a hypothesis as support for the hypothesis. Instead, biologists using adaptation to support proxy-based inferences use reasoning well described by Peirce's later model of abductive inference:

(1) The surprising fact, C, is observed;

(2) But if H were true, C would be a matter of course,

Hence,

(3) there is reason to suspect that H is true.[21]

The potential explanatory power of a hypothesis is reason to entertain and pursue a hypothesis and to seek its further evidence. But the evidence for the hypothesis must be independent of its explanatory power. I argue that we should carry this over to doing metaphysics.

The model of evidence used by biologists is a combination of two models: the vera causa ideal and robustness. The vera causa ideal holds that three things must be shown: the *existence* of a purported cause, the *competence* of the purported cause to produce the phenomenon, and the actual *responsibility* of the purported cause to the phenomenon (Herschel 1830/1987; Novick and Scholl 2017, 9). The existence and competence criteria must further be established independently of the responsibility by nonexplanatory reasoning. In proxy-based reasoning, so too must the correlation, functional adaptivity, and selection claims be established independently. I see my argument in part as an elaboration of the argument Rose Novick makes against using IBE in metaphysics on the grounds that biologists do not use it (Novick 2017; Paul 2012). At least, metaphysicians do not get to use it for free by mistakenly claiming it is used throughout science.

Robustness as evidence is the idea that detection of something by independent means is good evidence for the reality of the object. Wimsatt (1994) has argued for it as a criterion of reality. We see robustness used in the proxy-based reasoning as a way to ensure that proxies agree. Robustness can be used both to help establish a proxy (because the evidence for the other three claims is still inductive) and to calibrate the proxy relationship. Robustness can be seen as at odds with the vera causa ideal because robustness can be established using multiple independent lines of IBE. But the hypsodonty case shows that they are compatible.

4.2 Success

In the same way that I view the importation of a naively adaptationist just-so story method into metaphysics as worse than a waste of time, so too do I view the straightforward importation of the standard realist versus antirealist positions of scientific metaphysics and ontology. My hope is for the realism question to at least be changed in a way that allows new movement. Therefore, the success of a scientific practice sought cannot be empirical adequacy.[22]

One important part of this move is that, in adopting adaptation as the crucial idea upon which the proxy-based inference is built, the success of a scientific practice adapted to features of the world for purposes is a practical sense of success. The models of success here are the concrete success of successful control, successful intervention, successful creation, and successful prediction too.

The justification for practical successes is twofold. First, following Hacking (1983) and others, the emphasis should be on making and doing over thinking. I include prediction as doing though because the only reason for excluding it is an overreaction to overthinking. Second, practical successes are straightforward to identify and agree upon.

By using the variety of practical successes, it is also simple to keep in mind that the success of a practice is relative to a purpose. Just as traits of organisms are adapted for functions, so are traits of practices adapted for purposes. Even practices that are successful for multiple purposes are not therefore successful independent of any purpose.

We can raise the further question about how many concrete, practical successes we need in order to consider a scientific practice successful relative to a purpose. Speaking generally, we can only make the fallible claims that more are better than less and past successes are a sign of future successes.

4.3 Adaptation and Evolutionary Epistemology

My proposal here rings of evolutionary epistemology. I see my work here as differing in two main respects from the work done in evolutionary epistemology.[23] First, my emphasis is epistemological in a different way. Where evolutionary epistemologists aim at articulating how scientific change compares with biological change, my interest is in how the reasoning strategies used in philosophy of science differ from those used in biology.

Second, I do not see the need to wade into general arguments about whether adaptation in science should be seen as only analogous to adaptation in organisms or if both are kinds of a general adaptation. Just as alternative general philosophical positions can and should agree on basic methodological principles for practicing science (Sloep 1993), so too should different accounts of adaptation and selection on science agree on how to investigate and gain evidence for an adaptation hypothesis. What is needed is a kind of selection to operate on scientific practices, itself requiring variation, heredity, and fitness differences (Lewontin 1970; Godfrey-Smith 2007).

5. CONCLUDING REMARKS

This chapter proposes a new methodology for practicing metaphysics. To found a research program, we practice-based scientific metaphysicians must work out an exemplar to build on and modify going forward. Thankfully, rather than inventing a methodology de novo, we can co-opt and adapt the proxy by adaptation framework used by biologists to make inferences to unknown environments.

Whether the scientific and metaphysical problems should be regarded as analogous or rather the same but in different domains depends on the relationship between science and metaphysics. Considering my proposal at the meta-level, we are left with a picture of metaphysics as continuous with science on both epistemic and ontological dimensions. With its emphasis on inference, we can see this proposal in a long line of inferences based on a dualist ontology. For example, Locke asked how we can infer the properties of external objects from our experiences, leaving him open to strong Cartesian skepticism. However, my proposal is dualistic, not in a mind versus matter sense but in an organism and environment sense. A biological organism and its environment are mutually interacting and only pragmatically distinguished. While I have focused on learning about environments from organisms, we might also learn about organisms from environments.

Our picture of metaphysical knowledge is as with normal scientific inquiry: fallible but improving. The main difference is the dearth of data on scientific practice. We lack analyses, but more than that, we lack scientific diversity in the world. Metaphysicians can reason outward from ordinary experience much like how historical scientists reason backward from the present day using tools and methods developed using iteration and actualism. Skepticism is not in principle limiting to us; it is just that there are not enough possible data to know that much. But we will improve.

To conclude, I address three questions for this proposal. First, what kind of metaphysics can scientific practice inform us about? Second, what is the purpose of investigating metaphysical questions? And third, how does scientific metaphysics fit into the larger project of understanding how science works?

First then, what kind of metaphysics can scientific practice inform us about? As a methodological proposal for how to make strong inferences, it says nothing alone about the scope of metaphysics or scientific metaphysics. With the proxy by adaptation framework, we can only gain evidence for metaphysical claims that have some practical consequence for how scientific practice works. If this is the only methodology admitted, then practical consequences are the mark of knowable metaphysical truths. In this way, the proposal has an affinity with the pragmatist and positivist idea, uncontroversial within science, that the meaningful hypotheses are those that can be empirically verified.

What then is the scope of metaphysical claims that we can evidence with the proxy by adaptation framework? A better answer to this question will come from developing the metaphysical research program further, but some broad outlines can already be seen. On the one hand, first-order ontological claims such as "atoms exist" and "mushrooms are biological individuals" must make differences on scientific practices. If proxies for these kinds of claims cannot be developed, then the present proposal has nothing to offer here. On the other hand, metaphysical positions such as idealism and nominalism about the interpretation of first-order claims will probably not make differences to scientific practices. The proxy by adaptation framework is compatible with any metaphysical position that can make sense of first-order ontological claims and of selection operating on scientific practices. But it is unable to tell us about the natures of things that go beyond their practical effects in the same way that science is.

Second, what is the purpose of investigating metaphysical questions? I want to address whether the project of scientific metaphysics from biological practice is dualistic in the sense of Penelope Maddy's (2007) two-tier views. Maddy distinguishes between philosophers like Descartes for whom philosophical questions are continuous with and even arise from his scientific practice (one-tier view) and philosophers like Kant, Carnap, and van Fraassen for whom philosophical inquiry floats freely above science in asking questions about the interpretation and import of scientific fruits (two-tier views).

The present proposal fits neatly into neither one-tier nor two-tier views of philosophical inquiry as I understand them. It is like a two-tier view in seeming to imply that biologists themselves cannot say what is real in the biological world. It is like a one-tier view in using broadly the same forms of inquiry and asking similar questions as scientists—what is the structure of an organism, ecosystem, or whole domain?

Let us compare my proposal with the use of proxy-based reasoning from linguistic behavior. Anthropologist Caleb Everett and linguists Damian Blasí and Seán Roberts (2016) argue that the vowel structure of some spoken languages is adapted to the physical environment in which the language evolved and operates. In particular, they argue that desiccation of the environment will push languages against having a complex tonal system. They do not argue further for the tonality of a language as a proxy for humidity level in an environment, but it is possible that they could in the future.

Their article is the target article in the first volume and issue of the *Journal of Language Evolution*, with eleven commentaries on it. The commentaries are divided between internal methodological critiques of their analyses and external critiques about the very idea of language being adapted to the environment. The question I want to raise is whether these scientists should be characterized with a one-tier or two-tier view of linguistics. And my answer is that it isn't obvious that they fit into either neatly and further that the very question is of limited use. I see this question primarily about disciplinization because whether or not they are extra-linguist and asking meta-linguistic questions, their questions, inquiry, and subject matter are clearly scientific.[24]

Similarly then with scientific metaphysics from biological practice and biology, these are both different enterprises but both scientific. Further, they might interact the way that many distinct disciplines interact. The two-tier view of science and metaphysics is behind such quips as "philosophy is useful to science like ornithology is to birds." Metaphysics via a study of scientific practice is partially disconnected from the science. Scientists do not need to understand how science changes over time to conduct normal science, just as language users do not need to understand how language changes over time to have normal conversations. But metaphysics is continuous with science in that it comes from a scientific study of science as in linguistics.

Third, how does scientific metaphysics fit into the larger project of understanding how science works? Questions about scientific change and methodology are often characterized as epistemological and opposed to metaphysical questions about scientific realism because the two lines of thought are historically independent.

We should include how the world is in our analyses of scientific change and methodology. With a pragmatic approach to understanding scientific practice, the basic unit of analysis is: *Scientists* use *tools and methods* to investigate *aspects of the world* for particular *purposes*. This means that understanding scientific change and methodology is a relative significance problem (Beatty 1997) and we need to apportion relative responsibility to the following: scientists and the social structure of science, tools and methods, the world, and purposes. We should cite aspects of the parts of the world being investigated in explanations of why particular scientific practices work the ways that they do. Normal empirical features of the systems being studied influence how to investigate the system. Once we arrive at tentative metaphysical features of the domain being investigated by a scientific practice, we can use these features to explain why the practice is structured the way it is. This is the normal circularity of induction and deduction, and it is only vicious when the potential deductive strength is counted as inductive evidence as in IBE.

For example, Currie and Walsh (2018) have proposed an ontic-driven account of explanations of scientific methodology and used the differences between the motion of massive bodies and light to explain why Newton's methodology differs in mechanics and optics. A partially ontic account of scientific practice should be the norm, and whether a given practice is ontic- or socially driven is an empirical question. This problem is different from the standard realism versus anti-realism debate in being generally open to realism but considering realism problems about what exists and what it is like.

ACKNOWLEDGMENTS

Oliver Lean and I together came up with the initial idea for taking adaptation to be the relation between a successful practice and the world in Basel in August 2016.

I presented earlier versions of this chapter at the Central APA in St. Louis in 2016, to the *From Biological Practice to Scientific Metaphysics* Taiwan summer school in 2018, and to the Lake Geneva Biology Interest Group in 2019. I thank the audiences for their comments and questions.

I thank Ken Waters, Alan Love, Bill Wimsatt, Marcel Weber, and Ruey-Lin Chen for their feedback and encouragement.

I thank Mikael Fortelius and Indrė Žliobaitė especially for their comments on the use of teeth as an environmental proxy in section 2.

I have talked about this with just about everyone I know over the past seven years, and I thank you all for our conversations.

NOTES

1. Wallace's reasoning and evidence were refined by Darwin (1862) and many others, and conclusive proof that Angraecum sesquipedale is pollinated in the wild by the Xanthopan morganii praedicta actually only came in 1992 (Wasserthal 1997).

2. See MacCord 2017 for a history of morphological research on mammalian teeth.

3. See Currie 2018 for how a new species of platypus was reconstructed from a single tooth.

4. From personal communication with Fortelius, January 2019.

5. They first make a data model that groups the raw data: periodization into the MN system; tooth size into brachydont, mesodont, hypsodont; and so on. But they also use raw data for quantitative analyses. They also make decisions about which species and taxa to include in the analysis.

6. See the commentary that follows for my comment on the use of abduction versus inference to the best explanation and why I favor the former. Briefly, IBE takes the explanatory power of a hypothesis as support for the

hypothesis. Peircian abduction, however, takes the explanatory power of a hypothesis only as reason to pursue independent evidence for it.

7. Following Sober's (1984), "A is an adaptation for task T in population P if and only if A became prevalent in P because there was selection for A, where the selective advantage of A was due to the fact that A helped perform task T." Gould and Lewontin used an ahistorical definition of adaptation in Gould and Lewontin 1979. See Forber 2013 on different definitions of adaptation.

8. For an excellent analysis of how to study adaptation, see Olson and Arroyo-Santos 2015. These biologists focus on the different methods for studying adaptation: comparative, populational, and optimality. They also implicitly provide a strong defense of why IBE is not used by biologists here because of circularity.

9. Another solution is to find correlations using an independent proxy for the environmental condition being investigated. However, this pushes the problem back and makes the reasoning more complex. It also blurs the distinction I make between establishing a correlation and showing that the conclusion reached is robust.

10. Some such studies on other animals include the following: Schulz et al. 2013; Merceron et al. 2016; Müller et al. 2014; Karme et al. 2016; Winkler et al. 2019.

11. The sense of community here is just the species which live together and are at the same trophic level, a non-equilibrium concept of community.

12. "The main assumption is that there are already superior competitors in those environments which prevent the migrants establishing themselves" (Fortelius, personal communication, January 2019). They discuss aspects of this as macroevolutionary source-sink dynamics in a review paper (Fortelius et al. 2014), and they have modeled adaptation to harshening conditions in the context of the "species factory" (Fortelius et al. 2015).

13. Called "dental senescence," this has been shown in lemurs (King et al. 2005).

14. Specialization is another possible outcome of the selection pressure. Giraffes eating leaves high up in the trees are brachydonts (Fortelius, personal communication, January 2019).

15. I am not aware of these or any other scientists specially arguing for *ecological* adaptation via convergence. But not many biologists talk about ecological adaptation at all.

It would work in the analogous way though. A trait in two communities is probably the result of ecological selection because the communities both live in similar environments and are independent. The trait would be ecologically superior. You would need to find recent examples, before evolution could turn the ecological adaptation into an evolutionary adaptation.

16. Whether it is a convergence argument for using the proxy or a measurement of the proxy relationship is a tricky question that depends on the stage of inquiry, standing evidence, and purpose.

While epistemically philosophers would like proxies and indeed models and instruments to be justified before they are used to measure things, in practice these steps are not temporally sequential but operate in a feedback relationship.

17. This process happens for all scientific instruments. Chang (2004) describes it for thermometers. Van Fraassen (2010) discusses this in terms of the Problem of Coordination or how abstract signs represent what they represent.

18. I developed the following understanding through comments Waters has made informally. They should not be construed as his view.

19. The extent to which the practice-centered view takes us beyond theorizing, modeling, and experimenting is an open question. For many philosophers of science, it should do so, but the points are independent.

20. Gillet's distinction is related to Carnap's distinction between internal and external questions, but Gillet does not share Carnap's view that external questions are either meaningless or practical.

21. Peirce, "Lecture V," *Lectures on Pragmatism* (1903, CP 5.189). See Mcauliffe (2015) for how Peircian abduction got confused with IBE. Also, Peirce said a lot of things about abduction, and this formalization of it into inference form only captures some of it.

22. I thank Reuy-Lin Chen in particular for his comments on the importance of emphasizing practical success.

23. The most helpful survey of evolutionary epistemology for me was Renzi and Napolitano 2011.

24. Another example is the Forman thesis (Forman 1971) and its reception by historians of science. I see the idea that the Copenhagen interpretation of quantum mechanics ascended because of the environment in the Weimar Republic as a clear historical hypothesis. Some historians are upset by these kinds of hypotheses, as are some linguists. But if you accept that language and science are historically evolving, then they are valid hypotheses. The issue is evidence.

REFERENCES

Beatty, John. 1997. "Why Do Biologists Argue Like They Do?" *Philosophy of Science* 64 (Proceedings): S432–43.

Bender, Michael, Todd Sowers, and Edward Brook. 1997. "Gases in Ice Cores." *Proceedings of the National Academy of Sciences* 94, no. 16: 8343–49.

Campbell, Donald T., and Donald W. Fiske. 1959. "Convergent and Discriminant Validation by the Multitrait-Multimethod Matrix." *Psychological Bulletin* 56, no. 2: 81.

Chang, Hasok. 2004. *Inventing Temperature: Measurement and Scientific Progress.* Oxford: Oxford University Press.

Chang, Hasok. 2007. "Scientific Progress: Beyond Foundationalism and Coherentism 1." *Royal Institute of Philosophy Supplements* 61: 1–20.

Currie, Adrian. 2018. *Rock, Bone, and Ruin: An Optimist's Guide to the Historical Sciences.* Cambridge, Mass.: MIT Press.

Currie, Adrian, and Kirsten Walsh. 2018. "Newton on Islandworld: Ontic-Driven Explanations of Scientific Method." *Perspectives on Science* 26, no. 1: 119–56.

Damuth, J., M. Fortelius, P. Andrews, C. Badgley, E. A. Hadley, S. Hixon, C. M. Janis, R. H. Madden, K. Reed, and J. M. Smith. 2002. "Reconstructing Mean Annual Precipitation, Based on Mammalian Dental Morphology and Local Species Richness." *Journal of Vertebrate Paleontology* 22, no. 3: 48A.

Darwin, Charles. 1862. *On the Various Contrivances by Which British and Foreign Orchids Are Fertilised by Insects: And on the Good Effect of Intercrossing.* London: John Murray.

Eronen, J. T., Kai Puolamäki, L. Liu, K. Lintulaakso, J. Damuth, C. Janis, and M. Fortelius. 2010a. "Precipitation and Large Herbivorous Mammals I: Estimates from Present-Day Communities." *Evolutionary Ecology Research* 12, no. 2: 217–33.

Eronen, J. T., Kai Puolamäki, L. Liu, K. Lintulaakso, J. Damuth, C. Janis, and M. Fortelius. 2010b. "Precipitation and Large Herbivorous Mammals II: Application to Fossil Data." *Evolutionary Ecology Research* 12, no. 2: 235–48.

Eronen, Jussi T., P. David Polly, Marianne Fred, John Damuth, David C. Frank, Volker Mosbrugger, Christoph Scheidegger, Nils Chr Stenseth, and Mikael Fortelius. 2010. "Ecometrics: The Traits That Bind the Past and Present Together." *Integrative Zoology* 5, no. 2: 88–101.

Everett, Caleb, Damián E. Blasí, and Seán G. Roberts. 2016. "Language Evolution and Climate: The Case of Desiccation and Tone." *Journal of Language Evolution* 1, no. 1: 33–46.

Forber, Patrick. 2013. "Debating the Power and Scope of Adaptation." In *The Philosophy of Biology*, edited by Kostas Kampourakis, 145–60. Dordrecht: Springer.

Forman, Paul. 1971. "Weimar Culture, Causality, and Quantum Theory, 1918–1927: Adaptation by German Physicists and Mathematicians to a Hostile Intellectual Environment." *Historical Studies in the Physical Sciences* 3: 1–115.

Fortelius, Mikael. 1985. "Ungulate Cheek Teeth: Developmental, Functional, and Evolutionary Interrelations." *Acta Zoologica Fennica* 180: 1–76.

Fortelius, Mikael, Jussi Eronen, Jukka Jernvall, Liping Liu, Diana Pushkina, Juhani Rinne, Alexey Tesakov, Inesa Vislobokova, Zhaoqun Zhang, and Liping Zhou. 2002. "Fossil Mammals Resolve Regional Patterns of Eurasian Climate Change over 20 Million Years." *Evolutionary Ecology Research* 4, no. 7: 1005–16.

Fortelius, Mikael, Jussi Eronen, Liping Liu, Diana Pushkina, Alexey Tesakov, Inesa Vislobokova, and Zhaoqun Zhang. 2006. "Late Miocene and Pliocene Large Land Mammals and Climatic Changes in Eurasia." *Palaeogeography, Palaeoclimatology, Palaeoecology* 238, no. 1–4: 219–27. http://dx.doi.org/10.1016/j.palaeo.2006.03.042.

Fortelius, Mikael, Jussi T. Eronen, Ferhat Kaya, Hui Tang, Pasquale Raia, and Kai Puolamäki. 2014. "Evolution of Neogene Mammals in Eurasia: Environmental Forcing and Biotic Interactions." *Annual Review of Earth and Planetary Sciences* 42, no. 1: 579–604. https://doi.org/10.1146/annurev-earth-050212-124030.

Fortelius, Mikael, Stefan Geritz, Mats Gyllenberg, Pasquale Raia, and Jaakko Toivonen. 2015. "Modeling the Population-Level Processes of Biodiversity Gain and Loss at Geological Timescales." *The American Naturalist* 186, no. 6: 742–54. https://doi.org/10.1086/683660.

Gillet, Carl. Forthcoming. "Using Compositional Explanations to Understand Compositional Levels: Why Scientists Talk about 'Levels' in Human Physiology." In *Hierarchy and Levels of Organization in the Biological Sciences*, edited by Daniel Brooks, James DiFrisco and William C. Wimsatt. Cambridge, Mass.: MIT Press.

Godfrey-Smith, Peter. 2007. "Conditions for Evolution by Natural Selection." *The Journal of Philosophy* 104, no. 10: 489–516.

Gould, Stephen Jay, and Richard C. Lewontin. 1979. "The Spandrels of San Marco and the Panglossian Paradigm: A Critique of the Adaptationist Programme." *Proceedings of the Royal Society of London B: Biological Sciences* 205, no. 1161: 581–98.

Gould, Stephen Jay, and Elisabeth S. Vrba. 1982. "Exaptation—A Missing Term in the Science of Form." *Paleobiology* 8, no. 1: 4–15.

Hacking, Ian. 1983. *Representing and Intervening: Introductory Topics in the Philosophy of Natural Science.* Cambridge: Cambridge University Press.

Herschel, John Frederick William. 1830/1987. *A Preliminary Discourse on the Study of Natural Philosophy.* Chicago: University of Chicago Press.

Janis, Christine M., and Mikael Fortelius. 1988. "On the Means Whereby Mammals Achieve Increased Functional Durability of Their Dentitions, with Special Reference to Limiting Factors." *Biological Reviews* 63, no. 2: 197–230.

Jernvall, Jukka, and Mikael Fortelius. 2002. "Common Mammals Drive the Evolutionary Increase of Hypsodonty in the Neogene." *Nature* 417, no. 6888: 538–40.

Karme, Aleksis, Janina Rannikko, Aki Kallonen, Marcus Clauss, and Mikael Fortelius. 2016. "Mechanical Modelling of Tooth Wear." *Journal of the Royal Society Interface* 13, no. 120: 20160399.

King, Stephen J., Summer J. Arrigo-Nelson, Sharon T. Pochron, Gina M. Semprebon, Laurie R. Godfrey, Patricia C. Wright, and Jukka Jernvall. 2005. "Dental Senescence in a Long-Lived Primate Links Infant Survival to Rainfall." *Proceedings of the National Academy of Sciences of the United States of America* 102, no. 46: 16579–83. https://doi.org/10.1073/pnas.0508377102.

Kuhn, Thomas S. 1962. *The Structure of Scientific Revolutions.* Chicago: University of Chicago Press.

Lewontin, Richard C. 1970. "The Units of Selection." *Annual Review of Ecology and Systematics* 1, no. 1: 1–18.

Liu, Liping, Kai Puolamäki, Jussi T. Eronen, Majid M. Ataabadi, Elina Hernesniemi, and Mikael Fortelius. 2012. "Dental Functional Traits of Mammals Resolve Productivity in Terrestrial Ecosystems Past and Present." *Proceedings of the Royal Society B: Biological Sciences* 279, no. 1739: 2793–99.

MacCord, Katherine. 2017. "Development, Evolution, and Teeth: How We Came to Explain the Morphological Evolution of the Mammalian Dentition." PhD diss., History and Philosophy of Science, Arizona State University.

Maddy, Penelope. 2007. *Second Philosophy: A Naturalistic Method.* Oxford: Oxford University Press.

Mcauliffe, William H. B. 2015. "How Did Abduction Get Confused with Inference to the Best Explanation? " *Transactions of the Charles S. Peirce Society: A Quarterly Journal in American Philosophy* 51, no. 3: 300–319.

Merceron, Gildas, Anusha Ramdarshan, Cécile Blondel, Jean-Renaud Boisserie, Noël Brunetiere, Arthur Francisco, Denis Gautier, Xavier Milhet, Alice

Novello, and Dimitri Pret. 2016. "Untangling the Environmental from the Dietary: Dust Does Not Matter." *Proceedings of the Royal Society B: Biological Sciences* 283, no. 1838: 20161032.

Miller, Boaz. 2016. "What Is Hacking's Argument for Entity Realism?" *Synthese* 193, no. 3: 991–1006.

Müller, Jacqueline, Marcus Clauss, Daryl Codron, Ellen Schulz, Jürgen Hummel, Mikael Fortelius, Patrick Kircher, and Jean-Michel Hatt. 2014. "Growth and Wear of Incisor and Cheek Teeth in Domestic Rabbits (Oryctolagus Cuniculus) Fed Diets of Different Abrasiveness." *Journal of Experimental Zoology Part A: Ecological Genetics and Physiology* 321, no. 5: 283–98.

Novick, Aaron. 2017. "Metaphysics and the Vera Causa Ideal: The Nun's Priest's Tale." *Erkenntnis* 82, no. 5: 1161–76.

Novick, Aaron, and Raphael Scholl. 2017. "Presume It Not: True Causes in the Search for the Basis of Heredity." *The British Journal for the Philosophy of Science* 71, no. 1: 59–86.

Olson, Mark E., and Alfonso Arroyo-Santos. 2015. "How to Study Adaptation (and Why to Do It That Way)." *The Quarterly Review of Biology* 90, no. 2: 167–91.

Paul, Laurie A. 2012. "Metaphysics as Modeling: The Handmaiden's Tale." *Philosophical Studies* 160, no. 1: 1–29.

Peirce, Charles S. 1903. "Lecture V. " *Harvard Lectures on Pragmatism.* MS [R] 312.

Polly, P. David, Jussi T. Eronen, Marianne Fred, Gregory P. Dietl, Volker Mosbrugger, Christoph Scheidegger, David C. Frank, John Damuth, Nils C. Stenseth, and Mikael Fortelius. 2011. "History Matters: Ecometrics and Integrative Climate Change Biology." *Proceedings of the Royal Society B: Biological Sciences* 278: 1131–40. https://doi.org/10.1098/rspb.2010.2233.

Quine, Willard V. 1948. "On What There Is." *The Review of Metaphysics* 2, no. 5: 21–38.

Renzi, Barbara Gabriella, and Giulio Napolitano. 2011. *Evolutionary Analogies: Is the Process of Scientific Change Analogous to the Organic Change?* Newcastle upon Tyne: Cambridge Scholars Publishing.

Schulz, Ellen, Vanessa Piotrowski, Marcus Clauss, Marcus Mau, Gildas Merceron, and Thomas M. Kaiser. 2013. "Dietary Abrasiveness Is Associated with Variability of Microwear and Dental Surface Texture in Rabbits." *PLoS One* 8, no. 2: e56167.

Sloep, Peter B. 1993. "Methodology Revitalized?" *The British Journal for the Philosophy of Science* 44, no. 2: 231–49.

Sober, Elliott. 1984. *The Force of Selection: Evolutionary Theory in Philosophical Focus.* Chicago: University of Chicago Press.

Van Fraassen, Bas C. 2010. *Scientific Representation: Paradoxes of Perspective.* Oxford: Oxford University Press.

Vellend, Mark. 2010. "Conceptual Synthesis in Community Ecology." *The Quarterly Review of Biology* 85, no. 2: 183–206.

Wallace, Alfred Russel. 1867. "Creation by Law." *The Quarterly Journal of Science* 4, no. 16: 470–88.

Wasserthal, L. T. 1997. "The Pollinators of the Malagasy Star Orchids Angraecum Sesquipedale, A. Sororium and A. Compactum and the Evolution of Extremely Long Spurs by Pollinator Shift." *Plant Biology* 110, no. 5: 343–59.

Waters, C. Kenneth. 1994. "Genes Made Molecular." *Philosophy of Science* 61, no. 2: 163–85.

Waters, C. Kenneth. 2004. "What Was Classical Genetics?" *Studies in History and Philosophy of Science Part A* 35, no. 4: 783–809.

Waters, C. Kenneth. 2017. "No General Structure." In *Metaphysics and the Philosophy of Science: New Essays*, edited by Matthew Slater and Zanja Yudell. New York: Oxford University Press.

Waters, C. Kenneth. 2019. "Presidential Address PSA 2016: An Epistemology for Scientific Practice." *Philosophy of Science* 86, no. 4: 585–611.

Wimsatt, William C. 1994. "The Ontology of Complex Systems: Levels of Organization, Perspectives, and Causal Thickets." *Canadian Journal of Philosophy* 24, no. S1: 207–74.

Winkler, Daniela E., Ellen Schulz-Kornas, Thomas M. Kaiser, Annelies De Cuyper, Marcus Clauss, and Thomas Tütken. 2019. "Forage Silica and Water Content Control Dental Surface Texture in Guinea Pigs and Provide Implications for Dietary Reconstruction." *Proceedings of the National Academy of Sciences* 116, no. 4: 1325–30.

Žliobaitė, Indrė, Janne Rinne, Anikó B. Tóth, Michael Mechenich, Liping Liu, Anna K. Behrensmeyer, and Mikael Fortelius. 2016. "Herbivore Teeth Predict Climatic Limits in Kenyan Ecosystems." *Proceedings of the National Academy of Sciences* 113, no. 45: 12751–56.

3

WHAT WAS CARNAP REJECTING
WHEN HE REJECTED METAPHYSICS?

RICHARD CREATH

ON THE EVENING OF APRIL 6, 1922, in Paris, Henri Bergson confronted Albert Einstein on an issue of time. I will get into the details later, but the bottom line was that Bergson thought that by philosophic means he could show that Einstein's theory of relativity, especially the special theory of relativity, was importantly in error. Bergson spoke for twenty minutes; Einstein spoke for only one, and his reply included the seemingly undiplomatic sentence: "There is no philosopher's time."

The "debate" did not end there. Bergson produced a stream of books and papers, as did his students. Einstein fought back, too, but not as frequently. And the debate was not without consequences. Later that same year, Einstein won the Nobel Prize for physics—but not for relativity. He never did win a Nobel for his work on relativity. There were undoubtedly many reasons for the Nobel committee's decision, some better than others. But the controversy with Bergson was among them. In the award speech, the head of the committee, Svante Arrhenius, himself a Nobel Prize winner, said: "It will be no secret that the famous philosopher Bergson in Paris has challenged this [relativity] theory" (Canales 2015, 4). He went on to say that Bergson had shown that relativity "pertains to epistemology" rather than to physics. It was not philosophy's finest hour. Einstein's acceptance speech was about relativity rather than about the photoelectric effect, for which the prize was being awarded. One hundred years later, Einstein's theories of relativity are still doing well, and Einstein's reputation is secure. By contrast, few in the scientific community now take Bergson's views on physics seriously. That hardly shows that Einstein was right and Bergson wrong. But the longer-term success of

Einstein's views tends to disguise to us the fact that at the time real damage was done.

Rudolf Carnap rejected metaphysics utterly and completely, and Bergson was one of the few philosophers whom Carnap mentions by name as a metaphysician of the sort that he rejected (Carnap 1928/1967, 295; 1932/1959, 80). But Carnap says very little about what it was about Bergson to which he objected. I think Bergson's general view as well as this episode in particular are good examples of what Carnap meant by 'metaphysics.' Why? We'll have to look at Bergson in detail.

It may seem that the answer is obvious: Bergson was a metaphysician doing what metaphysicians do. This, however, is not enough of an answer. It gives the issue a name but provides no content. It seems informative because we know that metaphysics is a familiar branch of philosophy. And indeed, that is what the word 'metaphysics' means—to us. But however paradoxical it may seem, this is not what Carnap was rejecting when he rejected metaphysics, or so I shall argue.

This chapter is that argument. It is composed of three parts: First, I show that there is a significant interpretive problem in determining what Carnap's target was when he rejected metaphysics. That target cannot be what it is typically taken to be. Second, I consider a number of cases that are or might be thought to be metaphysics according to Carnap. Looking at these cases and gauging Carnap's reaction to them will help us articulate what Carnap means by 'metaphysics' and thus what his target was. I will pay particular attention to Bergson, specifically to the few remarks that Carnap makes specifically about Bergson and to the exchange in 1922 between Bergson and Einstein. These are particularly vivid examples of what Carnap means by 'metaphysics'. Finally, I will sketch a way out of the interpretive problem: a conception of what it is to do metaphysics (in the usual sense) without being metaphysical (in Carnap's sense). My thesis in all this is that we can identify what Carnap meant by 'metaphysics' when he rejected it. I do not intend to defend or justify Carnap's stance, only to identify its target.

1. THE INTERPRETIVE PROBLEM

As everyone knows, Carnap rejected metaphysics not just as false, sterile, or unknowable but as without cognitive meaning.[1] And it seems that he was rejecting what *we* mean by 'metaphysics'. And what is that? In common philosophical parlance, metaphysics is that branch of philosophy that treats of:

- The most basic features and relations of what is real
- Ontology
- Being qua being
- Necessity
- Such relations as part/whole and causation
- Such systems of relations as space and time

If we interpret Carnap as talking about all this when he rejected metaphysics, and I think that many contemporary readers assume that this is exactly what Carnap means, then we have a huge interpretive problem.[2] If we take Carnap to be rejecting the *field*, that is, the whole subdiscipline of philosophy that we call metaphysics, then we are forced to say that Carnap was rejecting what he himself did. He worked in this field.[3] He has an extended discussion of empirical reality *versus* metaphysical reality in the *Aufbau* (§§ 171–78). His "Empiricism, Semantics, and Ontology" (1950a) is one of the most important twentieth-century papers in the area, and it is still, some seventy years later, actively discussed (cf. Blatti and Lapointe 2016). His book *Meaning and Necessity* (1947) is only one of the many things he wrote on modality. His dissertation (1922/2019) was on space, a topic in which he maintained a lifelong interest. He wrote an introduction to Reichenbach's *The Philosophy of Space and Time* (1958), and his late philosophy of science book (1966) has an extended discussion of relativist space-time (especially pp. 144–76). There is no avoiding it; Carnap worked in the field of metaphysics and knew perfectly well that he did.

So, if we interpret "metaphysics" in Carnap's rejection of metaphysics as applying to the field as a whole, then we interpret Carnap as incoherent, that is, as rejecting what he himself is doing and moreover knows he is doing. He would be cutting off the limb on which he sits. Of course, there are those who are perfectly happy to interpret Carnap as babbling incoherently. This saves one the trouble of providing serious arguments against him. If no coherent alternative interpretation can be found, then we might have to rest with this interpretation. But we should not declare that there is no such interpretation of Carnap until we have looked for one.

2. CASES

In his well-known essay "Overcoming Metaphysics Through Logical Analysis of Language," Carnap (1932/1959) more or less takes it for granted that his readers understand the term *metaphysics* in the same way he does, so he does not clarify the word.[4] In a remark appended to the English translation of 1959, he is more explicit:

> *To section 1, "metaphysics."* This term is used in this paper, as usually in Europe, for the field of alleged knowledge of the essence of things which transcends the realm of empirically founded, inductive science. Metaphysics in this sense includes systems like those of Fichte, Schelling, Hegel, Bergson, Heidegger. But it does not include endeavors towards a synthesis and generalization of the results of the various sciences. (Carnap 1932/1959, 80)

Even this needs fleshing out. One way of explicating his usage is to see what he says about specific cases and to see what arguments he uses about either these cases or metaphysics in general. This is the strategy that Carnap himself recommends (cf. 1950b, 37). So in this section of the chapter, we will examine a couple of cases that Carnap definitely thinks of as metaphysical to see both what he says and why he might think of them as objectionable. As a contrast, we will also examine a couple of superficially similar cases that he specifically did not see as metaphysical and why he did not. We will also look at some of Carnap's further general remarks about metaphysics. This should yield both a better, that is, more faithful, interpretation of what Carnap meant by 'metaphysics' and also some understanding of why Carnap rejected metaphysics in his sense.

2.1 Mortimer J. Adler

I have already mentioned that Carnap took Bergson to be a prime example of a metaphysician in his sense. I will come back to Bergson in a moment. But first I consider a different case that Carnap discussed, Mortimer J. Adler, because Carnap is explicit there about his objections. Adler was a colleague of Carnap's at the University of Chicago and later became the editor of *The Great Books of the Western World* and of the *Encyclopedia Britannica* and also a mainstay of the Aspen Institute. People will differ in their assessments of his success in these endeavors. Here I look at Carnap's report (1963, 42) of a philosophy department seminar lecture by Adler and also at another lecture by Adler (Adler 1941).

Carnap does not give the date of Adler's department seminar lecture but says this about it:

Adler . . . declared that he could demonstrate on the basis of purely metaphysical principles the impossibility of man's descent from "brute," i.e.,
subhuman forms of animals. I had of course no objection to someone's
challenging a widely accepted scientific theory. What I found startling
was rather the kinds of arguments used. They were claimed to provide
with complete certainty an answer to the question of the validity or inva-
lidity of a biological theory, without making this answer dependent on
those observable facts in biology and paleontology, which are regarded
by scientists as relevant and decisive for the theory in question. (Carnap
1963, 42)

Plainly, Carnap is objecting to a conception of philosophy rather than to
a branch of philosophy. That way of philosophizing involves making a
priori claims about the world that purport to be substantive—in other
words, not about the language used or to be used—and to which the empirical science must conform.

The other lecture by Adler is "God and the Professors" of 1940, and it
shows a view much like that criticized by Carnap in the previous quote.[5] The
paper is a Jeremiad against almost all professors, whether of science or of philosophy, in American academia. He says that almost all of these professors are
"positivists" and goes on to hint at views that are like Carnap's or are carica-
tures of them. Once again, Adler proceeds on an a priori basis to assert the
following:

1. "Philosophy is superior to science . . ." because philosophy is ". . .
   knowledge of the being of things whereas science studies only their
   phenomenal manifestations . . ." (1941, 129).
2. "There are no systems of philosophy" only the one true one (1941, 129).
3. "Sacred theology is superior to philosophy . . . because it is more perfect
   knowledge of God and His creatures . . ." (1941, 131).
4. "Just as there are not systems of philosophy, . . . there is only one true
   religion, less or more embodied in the existing diversity of creeds"
   (1941, 131).
5. "Because God is its cause, faith is more certain than knowledge resulting
   from the purely natural action of the human faculties" (1941, 130).
6. "Science, philosophy, and theology cannot really disagree because they
   have different subject matters" (1941, 128–31).[6]

At the end of the same lecture, Adler (1941, 137–38) welcomes Hitler (this was 1940!) to cleanse America's universities of their professors. Earlier he had spoken of "liquidating" the professors, scientists and philosophers alike (Adler 1941, 134).

This lecture, "God and the Professors," clearly illustrates the features that Carnap found objectionable in the department seminar: a priori arguments that supposedly can give substantive results that overturn empirically established theories. If taken seriously, this can harm scientific progress. But I know of no evidence that Adler was taken seriously by anyone in the scientific community. He did have friends in high places, such as University of Chicago president Robert M. Hutchins and the publisher of *Time Magazine*, Henry Luce. Adler had a certain popular following as well. Certainly, inviting Hitler's armies to make a clean sweep in American academia can hardly be considered science-friendly. Moreover, Adler's claim that there are not many different systems of philosophy, but only the one true one, hardly squares with the history of philosophy, where there certainly seem to be multiple such systems, all impervious to his arguments. It is precisely such controversies that Carnap sought to sidestep in rejecting Adler's metaphysical approach and by adopting the principle of tolerance.

2.2 Henri Bergson

Henri Bergson is a vastly more influential and subtler philosopher than Mortimer Adler. And unlike Adler, Bergson was interested in what scientists had to say and was, in turn, taken seriously by some in the scientific community, at least at the time. He was also prolific and difficult to analyze. So it is not really possible to look at the full breadth of Bergson's views.[7] What I can do is highlight a few general threads of those views to which Carnap might and did object. I will then go on to look at two more particular arguments that Bergson raised against Einstein. My aim in all this is not so much to assess the merits of Bergson's views as to use his example to illuminate what Carnap means by 'metaphysics'.

Bergson was, at least until the First World War, wildly popular, both in academic circles and outside them. It was thought that perhaps only the Paris Opera could hold the throngs of people who wanted to hear him speak (Gunter 1969, 16). So far, we have seen only that Bergson had some sort of clash with Einstein. Since I think Bergson's texts are often unclear, let's begin by looking at a brief passage from his *An Introduction to Metaphysics*:

Now it is easy to see that the ordinary function of positive science is analysis. Positive science works, then, above all with symbols. Even the most concrete of the natural sciences, those concerned with life, confine themselves to the visible form of living things, their organs, and anatomical elements. They make comparisons between these forms, they reduce the more complex to the more simple; in short, they study the workings of life in what is, so to speak, only its visual symbol. If there is any means of possessing a reality absolutely instead of knowing it relatively, of placing oneself within it instead of looking at it from the outside points of view, of having the intuition instead of making the analysis; in short, of seizing it without any expression or symbolic representation—metaphysics is that means. *Metaphysics, then, is the science which claims to dispense with symbols.* (Bergson 1903/2012, 8–9, italics in the original)

Carnap actually quoted from this passage in saying that he, Carnap, is using the term 'metaphysics' just as many who claim to be metaphysicians use it:

> Other philosophers use the name "metaphysics" for the result of a nonrational, purely intuitive process; this seems to be the more appropriate usage:
>
> REFERENCES. In referring metaphysics to the area of the nonrational, we are in agreement with many metaphysicians. Cf., for example, Bergson ([Metaphysik] 5): "That science that wants to get by without symbols." This means that metaphysics does not wish to grasp its object by proceeding with concepts, which are symbols, but immediately through intuition. (Carnap 1928/1967, 295)

In identifying the nonconceptual with the nonrational, it may be that Carnap misunderstands the passage from Bergson that he has quoted.[8] But Carnap is saying that he and Bergson are in agreement about what metaphysics is, that they are talking about the same approach to philosophizing (though of course they take different attitudes toward it). In any case, my concern in this chapter is with what Carnap understood himself to be rejecting, so how he saw the matter is what is relevant.

What Bergson suggests is that the mind is divided into two parts. The first is the rational, intellectual, conceptual, analytical side. This is the side where we find ordinary scientific theories and the ordinary empirical observations on which science is based. The second side of the mind is the nonrational/nonconceptual part. This is the home of philosophical intuition, which is to be a direct and better grasp of how things really are. As we shall see, Bergson holds that the first part of the mind, including science, inevitably distorts that, but philosophical intuition can correct that.

This is not just my (or Carnap's) interpretation of Bergson. One of his more influential defenders, P. A. Y. Gunter (1969, 3–42 and esp. 29ff), says that Bergson holds that science is the product of the analytical intellect, which necessarily distorts reality.[9] Philosophy, however, has its source in the philosophical intuition, namely philosophical insight that is not part of the conceptual world of the intellect. This intuition can correct the distortions of the scientific intellect. As Gunter (1969, 29) puts it, "The intellect, especially the scientific intellect, is for Bergson a pragmatic faculty that, rather than comprehending things, utilizes them, and in utilizing them spatializes, fragments, and materializes them beyond recognition." Later, on the same page, Gunter rephrases this—four times: "intellectual analyses distort reality," "intellectual analysis distorts," "the intellect in most respects fragments, spatializes, and distorts reality," and "intellectual analysis spatializes and distorts reality."

This idea that science distorts, and it is the science of Newton or Darwin or Einstein that is referred to here by the phrase "intellectual analysis," is not the idea that scientific claims are fallible because they can be challenged and replaced by further applications of scientific methods. It is, rather, the idea that science as a whole distorts, even in the long run. These distortions could only be discovered as such because there is a source of knowledge that does not distort and that is thus in a position to correct the errors of science. This source is philosophical intuition, a nonconceptual direct apprehension of the nature of reality.[10]

One persistent theme in Bergson is the visceral conviction that scientific theories and all purely intellectual works are bloodless, abstract things, but the real world pulses with life. This is why the conceptual side of the mind must distort. Bergson's charge that the intellect/science distorts because it is abstract, however, misunderstands what science is supposed to do. Science is not supposed to reproduce the lived experience we have or to reproduce the world more generally. It is supposed to map it, to describe it, correctly one hopes. To say that the map does not turn green in the spring

(Nelson Goodman's phrase) is not to say that the map is incorrect or that the description is thereby a distortion (cf. Goodman 1963, esp. 552–54).

Bergson's antidote to the distortions of science is philosophical intuition. The word 'intuition' has had a long and tortured history in philosophy. In the seventeenth and eighteenth centuries, ordinary observation was called intuition. This was a representation of individuals compared for sameness and difference. In this sense, Kant has the mind turn a manifold of intuitions into an intuition of a manifold. There is also a long tradition of platonic intuition to underwrite mathematics and logic. Gödel appeals to this, as do Russell and Moore. This is usually distinguished from sensory evidence because its objects are not individuals and not in the causal order—in other words, they cannot be observed, even in principle.

Carnap spoke of intuition in his own work only very early (when he is a neo-Kantian discussing visual space) or very late (when intuitions are simply grist for linguistic explications). Carnap would have sharply distinguished Bergson's philosophic intuitions from the ordinary observations on which science relies, as does Bergson. Scientific observation, and more precisely observational judgments, are parts of the rational/symbolic/conceptual universe. And they are probative in science only because they are part of that domain. Bergson's philosophic intuition by contrast has its roots in the nonrational/nonconceptual mind. Carnap can accept that philosophical intuition is "experience" in a suitably broad sense, like a feeling of ennui. He would say it is a subjective attitude toward life. But in this sense, it is not a judgment at all and not one that can help select between two purely descriptive (nonevaluative) accounts of the world. A judgment *that* one is feeling ennui, however, is a perfectly ordinary observation, albeit about the mind. Carnap from day one is interested in intersubjective science and scientific observation, even in the *Aufbau* and before. The track record of intuitions, philosophic and platonic, shows them to be highly subjective and unable to resolve the apparent disagreements that inevitably arise among alternative intuitions. These irresolvable disagreements yield the "wearisome controversies" in philosophy that it is the whole purpose of Carnap's mature philosophy (from 1932 onward) to sidestep.

We have looked so far at Bergson's general view. Now we can turn to look more closely at two general objections that Bergson lodges against Einstein. First, Bergson claims that to treat time as a fourth dimension like that of the three of space is to "spatialize" time by positing a static, block universe in which nothing ever changes. Einstein is trying to "stop time"—in other words, to bridle the vital, creative forces that are beyond the reach of science. It is hard to see what Bergson even could mean by 'spatialize' or what defect is involved. It is likely that Bergson's claim that according to relativity theory nothing ever changes derives not from Einstein's treating time as a fourth dimension but rather from his treating it as being composed of points. There is a popular, though I think mistaken, understanding of Zeno's arrow paradox, according to which at every point, the arrow is not moving within that point.[11] Hence, the arrow is not moving at all. The combination of Zeno's idea with the idea that time is a series of points would explain the conclusion that in Einstein's theory nothing ever changes. Bergson's claim is not clear enough for us to be certain, but it is hard to see what else would explain his conclusion. Of course, the same criticism could be raised just as easily against Newton's treatment of instantaneous velocity and acceleration. It is hard to see how the objection has any force at all, but Bergson concludes that the basic entities should be processes rather than punctiform events. Moreover, time embodies the vital/living forces that permeate what is real and important and can never be captured by science—because they are creative rather than subject to laws.

What in all this would Carnap object to? Not to Bergson's process ontology. If scientists want to use a process ontology, that's fine. If philosophers want to develop rigorously what a process ontology involves, that's fine too. Does Carnap object that someone might disagree with a scientific theory? No. Does he object to Bergson's claims that for Einstein nothing ever changes and that Einstein is trying to "stop time"? Carnap would certainly think that these claims are based on misunderstandings of Einstein, but this is not what makes them metaphysics. Would Carnap object to Bergson's vitalism? Again, Carnap would not object to this if it were developed as an empirical theory or even as a well-worked-out conceptual framework, or at least he would not call either of these approaches to vitalism metaphysics.

Second, Bergson also objects specifically to the relativity of simultaneity embedded in the special theory of relativity. One might argue that present events are real in a way that future events are not yet real and that past events are no longer real. In this sequence of events, then, the "truly real" defines an objective simultaneity class. So Einstein's physics must be at best about what we know rather than about what is objectively real. This argument about the objectivity of the "now" is not a good argument, though there is some evidence (Carnap 1963, 37f.) that Einstein was bothered by it (not convinced, but bothered).

But this is not the argument that Bergson gave in the 1922 exchange with Einstein. Bergson's prose there is such that it is hard to see exactly what his argument is supposed to be. But he seems to argue that the relativity of simultaneity can be disproved on the basis of perfectly ordinary observations as follows:

1. I can have one experience of two nearby events such as a pair of flashes of light or a pair of notes, call them A and B.
2. Because this one experience is composed of experiences of A and B, my experiences of A and of B are simultaneous.
3. I can represent A and B as absolutely simultaneous, that is, as simultaneous independently of any inertial frame to which the events are measured. (A and B are taken as close to one another spatially but not exactly in the same spatial location.)
4. Therefore, the physical events A and B are absolutely simultaneous.
5. We can imagine a sequence of living beings observing a series of events, each near to the next, such that
   a. I observe A and B to be absolutely simultaneous, as noted.
   b. The second conscious being (Bergson suggests a "scientific microbe") observes B and C to be absolutely simultaneous.
   c. The third conscious being (microbe) observes C and D to be absolutely simultaneous.
   d. And so on.
6. We can establish on this observational basis that events, however distant from one another, are absolutely simultaneous.

Note that none of the claims here is nonconceptual because claims cannot be nonconceptual. The only way to be wholly nonconceptual is to say nothing. But premises 1 and 3 are about first-person reports. Bergson may well believe that such reports can be underwritten by philosophical intuition. While there are many issues surrounding first-person reports, let's ignore those issues and just grant these premises.

Claims 1–3 are also psychological, that is about mental states—mine. But 4 is about a physical state of affairs. Both Einstein and the third symposiast that day, Henri Piéron, challenged the legitimacy of such an inference in this case. Einstein began by noting the "the philosopher's"—in other words, Bergson's—concept of time is at once both psychological and physical. The former is about perceived time, while the latter is about events that are independent of us. Einstein took for granted that for perceived time there are no inertial frameworks to which temporal location and simultaneity even could be relativized. Thus, among psychological events there is no distinction to be drawn between absolute and relativized simultaneity. For physical events, however, there are alternative inertial frameworks. And relativity theory says that determinations of temporal location and simultaneity must be relativized to such frameworks. Einstein granted that drawing a temporal inference from the psychological to the physical often yields no conflict with the evidence, at least not in ordinary cases and for practical purposes, because, given the high velocity of light, the difference in temporal location from one inertial framework to another is too small to be observed. (As we shall see in a bit, this undermines Bergson's response in two ways.) Nonetheless, there are such differences. And no psychological evidence permits the inference from psychological simultaneity to absolute physical simultaneity even for events that are near to each other.

Einstein (1922/1969, 133) concluded: "Hence, there is no philosopher's time; there is only a psychological time different from the time of the physicist." Given the gloss after the semicolon, what appears before it means only that there is no concept of time such as the one that Bergson tries to employ such that an inference from claims about the mind to claims about absolute simultaneity can be warranted. Psychological and physical times should be kept distinct. This conclusion is not, as some have suggested, either "scandalous" or "incendiary" (Canales 2015, 5) any more than it would be for an eighteenth-century chemist to say that there is no philosopher's stone or for Lavoisier to say that there is no such thing as phlogiston.

Piéron was an empirical psychologist, and his comments on separating psychological and physical time are even more pointed. He gives essentially two arguments. First, he notes that his experimental results show that many factors other than physical temporal proximity can influence our perceptions of simultaneity. Hence any such inference is at best unreliable. Second, he notes that all observations are inexact, and so there can be no absolutely precise determinations of simultaneity even within the psychological realm. Piéron's (1922/1969, 134–35) conclusion is "Thus determinations of psychological succession or simultaneity can in no case be utilized as a measurement of physical time. . . . And thus the Bergsonian duration seems to me to be obliged to remain a stranger to physical time in general and in particular to Einsteinian time."

Bergson (1922/1969, 135) replied very briefly that he completely agreed that "the psychological establishing of a simultaneity is necessarily imprecise." He went on to add that, nonetheless, the psychological determinations are basic, that is, presupposed, by any instrument reading.

Bergson's concession that determinations of simultaneity are necessarily imprecise is fatal to the argument he gave against Einstein regardless of whether one is talking about psychological or physical simultaneity or about absolute or relativized simultaneity. The inference from the series of claims 5 to 6 has some chance of success if the simultaneities are exact because that relation is transitive. But the relation of *almost* simultaneity is not.

Bergson's remark that psychological determinations are presupposed in any instrument reading can be doubted,[12] but it does nothing to help his case, even if it is true. The inference from those psychological claims to Bergson's desired conclusion requires far more than we are given here. The inference is not justified by pure logic. Nor is it warranted by the scientific facts. And Bergson does not intend it to be a mere linguistic choice. In fact, it seems doubtful that Bergson intends any of these alternatives. There seems to be little left but to suppose that Bergson believes the inference to claim 4 and beyond to be justified by some sort of philosophical intuition that sees deeper and corrects what physics has to say. Moreover, it is fair to say that Bergson is not trying to formulate an alternative theory to deal with the evidence that Einstein and other physicists take to be relevant to deciding whether to accept his relativity theory. Rather, Bergson is approaching the issue from the outside.

So what in Bergson's argument would Carnap object to? Carnap would not object that it begins with first-person reports about experience or that Bergson wanted to use concepts other than those that Einstein wanted to use. If Bergson wants to use different concepts to formulate his own theory, that's fine. That would not be to transcend science via philosophical intuition, but to do science, alternative science, in the familiar way. Carnap would object that the argument is a bad one, that Bergson has not understood Einstein or the argument as given ignores the approximative character of the observation of physical events. Carnap would object, but these features are not what makes the enterprise metaphysical. What makes it metaphysics is that Bergson believes that his nonconceptual (and thus to Carnap nonrational) philosophical intuition can see deeper than or behind what ordinary science can see and thus be in a position to overrule it. Bergson attempts to force science to turn away from the concepts and patterns of inference that it finds most helpful in organizing experience and in organizing our response to it. This attempt is what makes it metaphysics. Such a mode of philosophizing impedes scientific progress. And insofar as intuitive insights are used on behalf of conflicting philosophical claims, we lack a way to resolve the issues, and the result is endless controversies.

2.3 Cases That Carnap Did Not Call Metaphysical

We have seen two cases that Carnap did identify as metaphysics. There are other cases where one might expect Carnap to reach the same judgment but where he does not. These can be treated comparatively briefly. Hans Reichenbach was a vigorous supporter of scientific realism, the idea that the unobservable entities that science postulates, such as atoms and electrons, are really there and not just convenient fictions. And during much of his career, ontology was a central focus for W. V. Quine. Ontology is a central part of the field of metaphysics, and yet Carnap denies that either of these men are metaphysicians. Why? The answer is that both are trying to turn their ontological claims into empirical ones. Perhaps this is obvious in the case of Quine, who rejected a priori methods altogether and insisted that even logic and mathematics were empirical. Reichenbach didn't go quite that far, but it did seem that he was defending scientific realism as an empirical claim. Certainly, Carnap (1963, 870) thought so.

This accords well with Carnap's comments describing metaphysics as an attempt to gain knowledge that somehow transcends the knowledge that empirical science can aspire to. Having such deeper knowledge, metaphysics would be able, from some philosophical or intuitive perspective, to "correct" the results of empirical science. It is this that Carnap rejects.

3. HOW TO DO METAPHYSICS WITHOUT BEING METAPHYSICAL

So is scientific metaphysics possible? As so often with Carnap, the answer depends on what one means, in this case by "metaphysics." If you mean what Carnap did by that word, namely an attempt from outside science to get at a reality that is behind or deeper than the results of ordinary empirical science, then the answer is of course not! But if you mean by "metaphysics" a branch of philosophy, rather than a way of going about it, then there is at least the possibility of working in this area in a scientific way.[13]

Carnap and his friends often spoke of "scientific philosophy," and whether philosophy is scientific according to Carnap's standards depends on how it is conceived and practiced. As we have seen, what Carnap was rejecting was plainly not a branch of philosophy as such, but a particular conception of philosophy. What he wanted was not philosophy as usual, or at least what was usual in early- and mid-twentieth-century Europe. Carnap did not want to *eliminate* or *overcome* or *uproot* metaphysics as a discipline or to do any of these things to philosophy as a whole. *Instead, he wanted to transform philosophy* so that it was not metaphysical in his sense, that is, so that it no longer tried to transcend science, that is, to reach deeper or higher knowledge than empirical science ever could. And he had an idea about how to do this.

Carnap wanted to reconceive the philosophical enterprise away from making (what purport to be) substantive claims about the world and that are claimed to be warranted by philosophical intuition. Instead, Carnap suggested that we think of the philosophical enterprise as one of making proposals for structuring the language of science. These proposals are not theories, not even tentative ones. They do not describe the world but have a different role. There is no fact of the matter about which of these proposals is the correct one because a language is not the sort of thing that is true or false. Philosophers can certainly engage in the highly useful task of exploring these linguistic structures to see how they work. This is a kind of conceptual engineering, but it is not describing the world around us. Philosophers, acting as amateur scientists, are free to describe the world around us. That is a perfectly worthwhile enterprise as long as it is evaluated on an empirical basis. And philosophical intuition is not any part of that empirical basis. But while these empirical descriptions are genuinely substantive/contentful, they are for the most part best left to one or more of the empirical sciences.

One is then free to adopt whatever language one wants, and in particular scientists are free to choose whatever language they find useful. Philosophers have no right to overrule them. For example, philosophers have no right to tell a Newton or an Einstein that they cannot use a language of points in space or instants in time or to define instantaneous velocity or instantaneous acceleration. Philosophers have no right to insist baldly that biologists must use a notion of kind or species of organism according to which species have essences that cannot change. Of course, it is permissible to challenge the logical consistency of an empirical claim or to defend such a claim against such a challenge. But such a discussion requires that the parties get clear about what the language of that empirical claim is by explicating the rules that structure that language. Moreover, it is permissible for philosophers to explicate terms in scientific discourse that they find unclear. Such terms can be clarified/precisified/explicated in multiple ways. What is not permissible is for a philosopher to insist that scientists must use some specific clarification. What this reorientation of philosophy does, then, is to give empirical science the primary role rather than treating philosophy as the queen of the sciences and capable of ruling them.

The reorientation of philosophy described here is itself a proposal on Carnap's part for how to talk about philosophy and science. It is designed to give philosophy a useful and important role in the overall scientific enterprise (a role not unlike that of mathematics) while at the same time sidestepping the "wearisome controversies" that seem to be the inevitable result of dueling philosophical intuitions. This reorientation is neatly summarized by Carnap in his principle of tolerance. It is the centerpiece of his mature philosophy.

> *In logic there are no morals.* Everyone is at liberty to build up his own logic, his own form of language, as he wishes. All that is required of him is that, if he wishes to discuss it, he must state his methods clearly, and give us syntactical rules instead of philosophical arguments. (Carnap 1934/1937, 52)

In saying that "there are no morals," he is saying only that no one language is the uniquely correct one. Some languages may be more useful than others, but scientists are free to make their own pragmatic choices as to which language they will use. As the years went by, the syntactic rules were broadened to include semantic rules as well. But the basic message remained the same. Philosophy is in the business of proposing and exploring conceptual structures—in other words, languages—that scientists may find useful. "*It is not our business to set up prohibitions . . .*" (Carnap 1934/1937, 51, italics in original). Scientists can talk of atoms and molecules or gross national products if these can be properly empirically grounded. And they are free not to use that language. They are free to use the language of real analysis or non-Euclidian geometries or four-dimensional manifolds of space-time points if that is helpful. And they are free to speak of selection pressures or to refuse to speak of species as having unchanging essences if that is useful in getting on with the business of describing the world. Scientists have the last word on what they need.

In this way, we can do metaphysics or any other branch of philosophy. We can explore realist or idealist languages. We can propose that we adopt a particular set theory. This sounds like metaphysics, and most now would call it that. We can propose a language in which 'Zero is a number' and 'Every number has a successor' are among the fundamental meaning-giving rules of the language. This sounds like ontology, and in one sense it is. Ontology as a field may be a branch of metaphysics construed as a subdiscipline of philosophy. But such proposals are not metaphysical in Carnap's sense. Nor would Carnap count proposals for a realist language—a language of physical things—or for a particular set theory as metaphysics in his sense. They are not what he is rejecting when he rejects metaphysics. As I said earlier, Carnap was plainly rejecting not a branch of philosophy but a particular conception of philosophy from whatever branch. He was out to transform philosophy, not to overcome it or any of its branches. And he was trying by example to show us how this might be done.[14]

NOTES

1. There is more to be said about what *cognitive* meaning is. But that is a topic for another essay. Carnap does not mean that the grammatically sound sentences involved express nothing whatsoever. They might very well express various attitudes. Rather the claim is that they do not succeed in describing the world truly or falsely.

2. See, for example, Bennett 2009, 38–39, especially footnote 2 (p. 38). The term *dismissivism* is Bennett's and is used by her as a "generic label for the view that there is *something* wrong with these debates" (p. 39): the ones that she has just described as "things that we metaphysicians think about" (p. 38).

3. This in no way denies that he worked in other fields as well.

4. In the available English translation, the title is "The Elimination of Metaphysics through the Logical Analysis of Language." "Overcoming Metaphysics through the Logical Analysis of Language" is closer to the sense of the original German.

5. This paper was originally published in a periodical that few academic libraries have: "God and the Professors," *Our Sunday Visitor, A Weekly Catholic National Newspaper,* December 1, 1940. It was reprinted in *Science, Philosophy and Religion: A Symposium,* New York, 1941, 120–38. Our references will be to the latter.

6. But neither here nor in the department seminar lecture does he seem to put any boundaries on the domain of either philosophy (he generally calls it "metaphysics") or theology.

7. Bergson's two most famous works relevant to this discussion are *Creative Evolution* (1911) and *Duration and Simultaneity: With Reference to Einstein's Theory* (1922/1965). As the title of the latter of these suggests, it is a defense of his position in the controversy with Einstein, discussed below.

8. Carnap's reading would seem to be plausible on its own. But it would have been particularly so to Carnap, given that the document from which Carnap's comment comes is the *Aufbau* (Carnap 1928/1967), which is now considered to be in many ways a neo-Kantian book. For Kant, the deployment of concepts is at the very heart of the rational enterprise, and to eschew concepts would be to eschew rationality itself.

9. Gunter (1969) has given a sympathetic and systematic interpretation of many of the major documents of and surrounding Bergson's controversy with Einstein, a reading that is still cited approvingly (Canales 2015, 237) by Bergson's contemporary advocates.

10. In saying this, Bergson is not necessarily anti-science, nor did he view himself as arguing against science. Apparently, he actively studied physics, biology, mathematics, and perhaps other sciences (Gunter 1969, 24, 29). But an interest in science hardly shows that he understood the individual scientific theories that he studied.

11. For an excellent treatment of this and related paradoxes and their significance for using a dense ordering for temporal elements, as in talk of instantaneous velocity or acceleration, see White 1992, 177–79.

12. Such claims are, of course, an article of faith in the "way of ideas" tradition stemming from Descartes. It was a popular view among both rationalists and empiricists for centuries and well into the twentieth century. I have no intention of challenging it here, but it is somewhat less popular now.

13. One prominent defender of what he and others call "scientific metaphysics" is C. Kenneth Waters. See especially his 2017 publication, where he says: "For the purposes of this chapter, I will assume that metaphysics is an area of philosophy that seeks to answer questions about the general nature of reality . . ." (83).

14. Thanks to Elliott Sober for first stimulating my interest in the Bergson–Einstein exchange. Thanks also to colleagues Steve Elliott, Jane Maienschein, and especially Ronald Hoy for comments on earlier versions of this chapter.

REFERENCES

Adler, Mortimer J. 1941. "God and the Professors." In *Science, Philosophy, and Religion,* 120–38. New York: Conference on Science, Philosophy, and Religion in Relation to the Democratic Way of Life, Inc.

Ayer, A. J., ed. 1959. *Logical Positivism.* New York: Macmillan Publishing.

Bennett, Karen. 2009. "Composition, Colocation, and Metaontology." In *Metametaphysics: New Essays on the Foundations of Ontology*, edited by David Chalmers, David Manley, and Ryan Wasserman, 38–76. Oxford: Clarendon Press.

Bergson, Henri. 1903/1912. *An Introduction to Metaphysics.* Translated by T. E. Hulme. New York: G. P. Putnam's Sons.

Bergson, Henri. 1911. *Creative Evolution.* Translated by Arthur Mitchell. New York: Henry Holt and Company.

Bergson, Henri. 1922/1965. *Duration and Simultaneity: With Reference to Einstein's Theory.* Translated by Leon Jacobson. Indianapolis: Bobbs-Merrill Company.

Bergson, Henri. 1922/1969. "Text of Bergson's Remarks in the 1922 Exchange with Einstein and Piéron." In *Bergson and the Evolution of Physics*, translated by P. A. Y. Gunter, 122–33, 135. Knoxville: University of Tennessee Press.

Blatti, Stephan, and Sandra Lapointe, eds. 2016. *Ontology after Carnap.* Oxford: Oxford University Press.

Canales, Jimena. 2015. *The Physicist and the Philosopher: Einstein, Bergson, and the Debate That Changed Our Understanding of Time.* Princeton, N.J.: Princeton University Press.

Carnap, Rudolf. 1922/2019. "Space." Translated by Michael Friedman et al. In *The Collected Works of Rudolf Carnap, Vol. 1, Early Writings*, edited by André Carus et al., 21–208. Oxford: Oxford University Press.

Carnap, Rudolf. 1928/1967. *The Logical Construction of the World.* Berkeley: University of California Press.

Carnap, Rudolf. 1932/1959. "The Elimination of Metaphysics Through Logical Analysis of Language." Translated by Arthur Pap. In *Logical Positivism*, edited by A. J. Ayer, 60–81. New York: Macmillan Publishing.

Carnap, Rudolf. 1934/1937. *The Logical Syntax of Language.* Translated by Amethe Smeaton. London: Kegan Paul Trench, Trubner & Co.

Carnap, Rudolf. 1947. *Meaning and Necessity.* Chicago: University of Chicago Press.

Carnap, Rudolf. 1950a. "Empiricism, Semantics, and Ontology." *Revue International De Philosophie* 4, no. 11: 20–40.

Carnap, Rudolf. 1950b. *Logical Foundations of Probability.* Chicago: University of Chicago Press.

Carnap, Rudolf. 1958. "Introductory Remarks to the English Edition." In *The Philosophy of Space and Time,* translated by Hans Reichenbach, Maria Reichenbach and John Freund, v–vii. New York: Dover Publications.

Carnap, Rudolf. 1959. "Remarks by the Author." In *Logical Positivism,* edited by A. J. Ayer, 80–81. New York: Macmillan Publishing.

Carnap, Rudolf. 1963. *The Philosophy of Rudolf Carnap,* edited by Paul Arthur Schilpp. LaSalle, Ill.: Open Court.

Carnap, Rudolf. 1966. *Philosophical Foundations of Physics: An Introduction to the Philosophy of Science.* New York: Basic Books.

Carnap, Rudolf. 2019. *The Collected Works of Rudolf Carnap, Vol. 1, Early Writings,* edited by André Carus et al. Oxford: Oxford University Press.

Chalmers, David, David Manley, and Ryan Wasserman, eds. 2009. *Metametaphysics: New Essays on the Foundations of Ontology.* Oxford: Clarendon Press.

Einstein, Albert. 1922/1969. "Text of Einstein's Remarks in the 1922 Exchange with Bergson and Piéron." In *Bergson and the Evolution of Physics,* translated by P. A. Y. Gunter, 133. Knoxville: University of Tennessee Press.

Goodman, Nelson. 1963. "The Significance of *Der Logische Aufbau.*" In *The Philosophy of Rudolf Carnap,* edited by Paul Arthur Schilpp, 545–58. LaSalle, Ill.: Open Court.

Gunter, P. A. Y. 1969. *Bergson and the Evolution of Physics.* Knoxville: University of Tennessee Press.

Piéron, Henri. 1922/1969. "Text of Piéron's Remarks in the 1922 Exchange with Bergson and Einstein." In *Bergson and the Evolution of Physics,* translated by P. A. Y. Gunter, 133–35. Knoxville: University of Tennessee Press.

Schilpp, Paul Arthur, ed. 1963. *The Philosophy of Rudolf Carnap.* LaSalle, Ill.: Open Court.

Slater, Mathew, and Zanja Yudell, eds. 2017. *Metaphysics of Philosophy of Science: New Essays.* Oxford: Oxford University Press.

Waters, C. Kenneth. 2017. "No General Structure." In *Metaphysics of Philosophy of Science: New Essays,* edited by Mathew Slater and Zanja Yudell, 81–107. Oxford: Oxford University Press.

White, Michael J. 1992. *The Continuous and the Discrete.* Oxford: Oxford University Press.

4

IDEAL OBSERVATIONS
Information and Causation in Biological Practice

OLIVER M. LEAN

1. INTRODUCTION

From the way biologists tend to talk, it seems like biology involves two very different types of phenomena simultaneously. DNA is used to produce proteins by complex molecular interactions, for example, yet in doing so it is also said to "code for" those proteins. When a neuron in the brain depolarizes, it's said to be passing along a "message." When a gazelle leaps high into the air, it is "signaling" to others around it. In all these cases, everyone agrees that there is a physical process unfolding according to causal principles of some kind. Yet at the same time, talk of coding and signals and messages suggests that these processes also involve *information*. References to information and related terms are common in biological sciences covering a range of scales, from the molecular level of genetics to ethology (the study of animal behavior) and neuroscience. *Information-talk*, as I will call it, is a common and widely accepted feature of the language of biologists.

What is less clear, however, is how to interpret this fact. Information-talk has a number of features that appear at first glance to be puzzling. For example, biological information seems to be intimately related to yet importantly different from the physical stuff of biological phenomena: A physical thing can be said to carry different pieces of information, and the same information can appear in different physical forms. Hence, information can be "transmitted" from one physical thing to another, potentially very different, thing. On its face, these are observations with *metaphysical* significance: What is this "information" biologists seem so comfortable talking about, and how does it fit into the wider scientific picture of the biological world? In particular, how does it relate to the relatively uncontroversial *causal* or *physical* aspects of biology? Is information-talk just a convenient shorthand for what are in fact garden-variety physical phenomena (Sarkar 2000; 2005)? Or does information play a deeper and more essential role that cannot be accounted for in purely physical terms (Barbieri 2016)?

Why does this matter? The idea of biological systems trading in information lies at the heart of various debates—some intrinsic to science and its methods, some with wider social significance. For some, the role of information is deeply important to what distinguishes living things from nonliving matter (Stotz 2019), perhaps even a *biosignature* that may help us recognize life elsewhere in the universe (Walker et al. 2018). For others, talk of information is a historical relic that confuses more than it clarifies and so should be dispensed with. Some ethologists, for instance, have argued that talk of information in animal signals is misleading because focusing on an intangible "content" to signals diverts attention from important physical aspects of signal design. Instead, they argue, we should understand animal communication as the attempt by senders to *manipulate* receivers (Rendall et al. 2009; Owren et al. 2010). Some even argue that information-talk should be resisted on political grounds: the idea that genes carry information, some believe, descends from outmoded and harmful ideas about the primacy of genetic factors over others in development—that certain traits are "genetically encoded" and hence immutable (Francis 2003).

What we have, then, is a question with a metaphysical flavor and that concerns the biological world and the study of it. In this chapter, I offer a way to address this question of how to interpret information-talk in biology. Importantly, my discussion goes further than an analysis of the logical or linguistic relationships between information and other concepts intrinsic to biological theory; it is not just an exercise in *metaphysics of science*, in the sense discussed in the introduction to this volume. Instead, it has implications for how the use of that concept drives the success of biological practices—specifically, practices of investigating and explaining how biological things solve what I will call *coordination problems*. Because of this, it qualifies as a work of *scientific metaphysics* in the sense outlined in the introduction to this volume and represented throughout other chapters. That is, it has implications for not just what biological theory is like but what the biological world is like.

I take the view (alongside other contributors to this volume) that such questions are best done with an explicit and well-defined purpose to motivate and constrain our theorizing. To that end, my particular aim here is to make sense of biological information in a way that suits what I henceforth call *naturalized epistemology*. The "naturalism" I refer to here is one that allows scientific developments to place constraints on other areas of our thinking. With that in mind, naturalized *epistemology* is the project of situating epistemic concepts—including *belief, knowledge,* and *meaning*—within a scientific picture of the world, potentially reshaping those concepts in order to make them fit. If we want to continue to understand ourselves as knowing subjects with beliefs about the world that can be true or false, a commitment to naturalism compels us to reckon seriously with science's picture of humans as physically embodied organisms evolved to solve physical, biological, and social problems.

Many take the concept of *information* to play a vital role in fitting these two pictures together. On the one hand, information as a concept has its origin in mentalistic talk: it invokes epistemic ideas like instruction or evidence. On the other hand, as seen previously, information appears to be commonplace in respectable scientific discourse about biological phenomena. Because information appears to stand astride these two realms, it's thought that we may be able to narrow the conceptual chasm between them by tracing the origins of "full-blown" mental phenomena to biological processes that are more generally informational (Dretske 1981; Millikan 2004; Skyrms 2010; Sterner 2014; Garson and Papineau 2019).

To achieve this, however, one must first lay the aforementioned metaphysical groundwork: we need to define exactly what is meant when we say that biological phenomena are trading in "information." To do the work of naturalizing mental or epistemic notions, this account of biological information should satisfy at least two closely related criteria.

First, the account must imply that biological systems are "really" informational in a literal or substantive sense (Collier 2008). In other words, information should not be merely something we project onto what are in fact plainly physical phenomena. To use a classic example, the rings of a tree are often said to carry information about its age, but this is apparently just information *for us,* the observers. Instead, what we need is a sense in which something is information *for the organism* itself, in some sense independently of us. Without this, we lack a basis for recognizing epistemic phenomena in the natural world because "information" remains confined to the minds of the biologists.

Secondly, we need a clear sense of how information so defined relates to the *causal* aspects of biological phenomena. Everyone agrees that causality is vitally important to our understanding of how biology works. The key question is what talk of "information" adds to that picture—what explanatory role it can play—that ordinary causal/physical language cannot do by itself. Some have addressed this challenge by effectively *reducing* the latter to the former—that is, explicating information *in causal terms*. Holly Andersen (2017), for instance, expresses information as a measure of pattern over a causal structure. Elsewhere, recent and widely discussed work by Griffiths and colleagues has analyzed biological information simply as a measure of *causal specificity*—of the extent to which a cause yields fine-grained or precise control over its effect (Griffiths et al. 2015; Stotz and Griffiths 2017; Stotz 2019; see section 3.3). Yet for the particular purposes of reconceiving *epistemic* phenomena in biological terms, this reductive approach seems at best incomplete: What does causal specificity have to do with knowledge and related concepts? It's proven difficult to answer that question in a way that naturalism would deem acceptable.

Taken together, these interwoven criteria establish the puzzle that an adequate characterization of information in biology should answer: What we need is a way of understanding biological information that is both (1) metaphysically integrated into the established scientific picture of biology and (2) playing a substantive explanatory role within that picture.

While there is still much debate about exactly how to solve this puzzle, there has emerged a broad (though far from universal) agreement about what the notion of biological information is broadly about; namely, that it has something to do with *use*. This general idea is represented in a wide variety of works in a wide variety of fields. For example, it is represented in the teleosemantic approach to representational content of Millikan (2004; 2013), Neander (1995; 2017), and others, which (very roughly) ties the content of a representation to the biological function of that representation's consumer. In short, something means what it has the function of meaning within a system evolved to solve adaptive problems. Variations on this general theme of information are also present in Dretske (1988), Maynard Smith (2000a; 2000b), Collier (2008), Skyrms (2010), Shea (2007; 2013), Bergstrom and Rosvall (2009), Anderson and Rosenberg (2008), Seyfarth et al. (2010), Robinson and Southgate (2010), Kight et al. (2013), and Lean (2014), to name just a few. It is also strongly represented in the ecological approach to cognitive science (Gibson 1966; 1979; Chemero 2003; Baggs and Chemero 2018), which treats the environment as containing information that is perceived by the organism as affordances for achieving its own biological aims. Similar appeal to use and purpose is at work in the context of cognitive science: Bechtel (2009), for one, argues that the proper conception of "information" is the one offered by control theory, which is concerned with the design of systems that exert control over factors affecting those systems by sensing and responding to changes in those factors. This covers an extremely diverse range of works, whose similarities and differences are too many and too diverse to discuss here. The key point at present is that they broadly agree on at least one thing: namely, they treat "information" as something that is being exploited by an agent (or something like an agent) to achieve some functional purpose.

I will call this general point of agreement the *use-consensus*, and this consensus will be the central focus of this chapter. As is sometimes but not always acknowledged, this is inherited directly or indirectly from Peirce's theory of signs and might be seen as an extension of Peirce's account of meaning to the living world in general (Short 2007). One way these views differ is in how exactly they understand the idea of function that fixes the meaning of a representation or even whether the "information" in question is best thought of as meaningful at all. I use the term *use-consensus* as an attempt to remain as neutral as possible about these issues—to capture what is shared between these approaches and to abstract from their differences. Rather than proposing yet another variation of this view, I offer a *reconstruction* of the use-consensus from the perspective of practice as an operational principle that organizes certain kinds of scientific investigation.

A key feature of my account is that it is closely modeled on Woodward's (2003; 2010) interventionist analysis of causation, which similarly aims to understand causal reasoning as a cognitive tool designed to serve various scientific aims. Ereshefsky and Reydon (2023) see Woodward's framework as similar in spirit to their grounded functionality account (GFA) of natural kinds: in their terms, reasoning about causes serves various scientific functions in ways that are grounded in the world. With this in mind, one might think of this chapter as something like a grounded functionality account of biological information—one that is explicitly compared and contrasted with Woodward's account of causal reasoning.

A benefit of this method of analysis, I hope to show, is that it brings to the foreground answers to the preceding conditions for a substantive account of biological information: First, it lays in sharp relief how this information is related to yet importantly distinct from the causal features of those phenomena. Specifically, it finds that causation and information need not be ontologically distinct in any strong sense; rather, they correspond to different ways of thinking about a given system for different purposes. Second, it shows when and why information can be said to be "literally" or substantively at work in biological phenomena. The answer it gives to this question embodies a form of naturalism in the spirit of Price (2011): since scientific observers are themselves living things, information "for us" is simply a special case of information "for the organism." In a sense, then, all information is biological information.

The chapter proceeds as follows. Section 2 outlines the interventionist account of causation as developed by Woodward, highlighting key features that I will borrow and adapt in order to explicate informational reasoning in similar terms. Section 3 then develops this framework for understanding informational reasoning: First, I characterize it in general terms that include non-biological cases of informational thinking using a simple case of scientific measurement. There I introduce the concept of an *ideal observation*, which makes sense of the informational aspects of a system and how these differ from both its causal and statistical features. Then I show how the same framework can be used to ground a substantive view of *biological* information. Section 4 concludes.

2. CAUSAL REASONING: CONTROL AND INTERVENTION

In this section, I outline key features of the interventionist account of causation, particularly the version defended by Woodward (2003; 2010; 2014)—henceforth *ICW*. As we will see, this framework is founded on explicitly pragmatic ideas: it aims to understand causality by asking what purposes the use of the concept serves in scientific practice rather than aiming to ground causality in a particular scientific or metaphysical theory. This focus on the function of causal reasoning, rather than on some intrinsic nature of causes as such, is embodied in the way it understands the commitments made by causal claims. It is this method of analysis that I aim to replicate in my analysis of the concept of information in the next section.

The first and most important feature of ICW is that it takes causality to be intimately tied to the notion of *control*: humans are not passive observers of our reality; we *act* in the world to effect changes, and our idea of causality guides reasoning about the differences our actions can make. Jenann Ismael (2017) expresses this motivation for interventionism in terms of affordances:[1] "To the embedded agent who doesn't just observe, but also intervenes in, his environment, the world is chock-full of opportunities and affordances. The terms in which he represents the world will be designed to disclose them. Causal relations are the generic form of these opportunities and affordances" (117).

The idea that causal reasoning reveals affordances for control is built into the way ICW analyzes claims about causal relationships. Consider three variables connected by edges, representing a physical system (Figure 4.1). These variables form a *directed acyclic graph* or DAG:

$$W \longrightarrow X \longrightarrow Y$$

Figure 4.1. A simple causal chain represented by a directed acyclic graph (DAG).

In the preceding graph, the edges connecting W and X and X and Y represent the idea that W causes X and X causes Y. (The language of familial relationships is useful here: W is a "parent" of X, and X and Y are "descendants" of W.) Interventionists in general hold that no amount of probabilistic or statistical information about these variables can express the idea that they are *causally* related: there is no way to express that X is a cause of Y, for instance, in terms of conditional dependencies between the variables. Instead, to say that X is a *cause* of Y is to say that Y would change if X were manipulated from outside the system—that *changing* X *would change* Y.

Hence, causality on this view is essentially tied to the idea of *intervention*. Pearl (2009), whose work informs ICW, characterizes interventions algebraically: an intervention on X "breaks" its upstream dependencies on its parents and turns it into an independent variable. In contrast, a peculiar feature of ICW in particular is that it characterizes interventions as themselves causal processes of a very particular kind, which Woodward calls *ideal interventions*: An ideal intervention on X with respect to Y is "a causal process that changes the value of X in an appropriately exogenous way, so that if a change in the value of Y occurs, it occurs only in virtue of the change in the value of X and not through some other causal route" (Woodward 2003, 94). Treating interventions as themselves causal processes makes it possible to explicitly represent them, as shown in Figure 4.2.

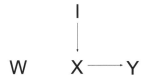

Figure 4.2. An intervention (represented by *I*) on *X* in the causal chain from Figure 4.1. In ICW, interventions are considered causal processes themselves, hence representable as such in DAGs.

Hence, for Woodward, to claim that X is a cause of Y is to claim that an ideal intervention on X changes Y, or at least the probability distribution over Y. Since the intervention detaches X from its causal dependency on its parent W, as seen in Figure 4.2, any remaining correlation between X and Y must be due to a causal relationship between them. This view of the meaning of causal claims aims to account for how scientists actually test and refine their causal models of the world (see Woodward 2014 for more detail).

As well as distinguishing causes from non-causes, this framework allows us to differentiate *between* causes of some effect along a number of different dimensions (Woodward 2010). For example, one cause variable is more *specific* than another if it exerts more fine-grained control over the value of the effect variable; that is, if it has a greater number of values that each specify a particular value of the effect. As mentioned previously, it is this *causal specificity* that Stotz and Griffiths (2017) claim underwrites the notion of informational relationships in biology—an issue to which I'll return in section 3.3.

ICW has met with several criticisms, the answers to which will be useful for understanding what follows. One class of objections centers around the view that it is overly anthropomorphic and subject-focused. We generally want to understand causality as something *objective* that actually governs nature independently of us and our beliefs about it, it is argued. Given that, it may seem that defining causality in terms of interventions implies that there are no causes in the natural world without agents who act on it. However, there is a way to understand how causal claims on ICW can be understood to be objective—that is, meaningful and true in a way that is, in an important sense, independent of the subject(s) making the claims. Since I will be adapting this aspect of ICW in my complementary analysis of information, I will take some time to address it.

The objectivity of causal claims in ICW comes from the fact that the interventions they posit are *idealized*. To idealize in scientific theorizing is to introduce assumptions that are false or unrealistic as a useful theoretical device. With that in mind, what false assumptions are made with respect to ideal interventions? Firstly, the ideal interventions central to ICW are essentially *counterfactual*: they posit differences in one and the same event if the value of a variable *were* different. This is an idealization because, by definition, it is not possible to realize and compare a set of mutually exclusive counterfactuals; instead, the best we can do is *approximate* this counterfactual experiment by repeating a process under controlled conditions in which only the (putative) cause is varied. Because they are counterfactual, then, ideal interventions are hypothetical (Woodward 2003, 40).

Secondly, ideal interventions are ideal in the sense that they are assumed to have an ideal degree of surgical precision: they are both causally and statistically unrelated to any other variable in the system. This is an idealizing assumption because it is always at least logically possible that there is a confounding factor that we have not controlled for or that our intervention had some other unintended effect. The best we can do, then, is to *approximate* these ideal interventions by controlling as much as possible. This account fits well with various aspects of scientific practice with respect to testing causal claims: For example, I've argued elsewhere (Lean 2020) that to maximize *binding specificity* in drug design is effectively to approximate an ideal intervention on the drug target.

It is these idealized features of interventions that render the associated causal claims objective in a specific but important sense. In particular, it makes sense of the idea that one thing can cause another whether or not we actually perform the kind of interventions necessary to test this claim. We can, for example, coherently talk about causal relationships in the deep past or between distant celestial bodies despite the fact that we can't go back in time and lack the power to divert the course of planets: when we discuss these causal relationships, we are discussing *hypothetical* interventions on the causes, not actual ones. For example, to say that a meteorite caused the extinction of the dinosaurs means that if, say, the meteorite had been diverted to miss Earth, the mass extinction would not have occurred. This has the important semantic consequence that the truth of causal claims is not relativized to the actual abilities and actions of agents: they are about manipulability *in principle*, not necessarily in practice.

In addition to making causal claims independent of our abilities, this also makes them independent of our beliefs: even if our beliefs about the causal structure of the world are based on well-executed controlled experiments, those beliefs can still be wrong. This is because the experiments we actually perform to test causal claims can only be approximations of the counterfactual experiments that those causal claims are about, and so actual experimental results do not logically entail the truth or falsity of causal claims. Of course, this disconnect between the *meaning* of causality and the empirical observations we take as *evidence* of causal connections is central to Hume's discussion about the concept. Nevertheless, causality is central to our reasoning about the world despite these worries (as Hume observed), and ICW provides a rich account of this central role as it manifests in scientific practice.

Broadly, then, the effect of idealizing the interventions posited in causal claims is that it removes from those interventions any explicit reference to agency or purposeful action. It is "heuristically useful," Woodward says, to think of an ideal intervention as the action of an agent; however, it is possible to characterize ideal interventions without endowing them explicitly with agency; instead, we can simply stipulate causal and statistical conditions. If an action is a relationship between an agent and the system on which they act, successful control depends on both. In idealizing the role of the agent in that relationship—that is, granting all the necessary statistical and causal features of the intervener—the absence of an effect can therefore be attributed to a lack of causal power in the target variable, not to any failure to intervene properly on it. This feature, in my view, strikes a careful balance between acknowledging the essential role of agency and purpose in causal reasoning while maintaining a sense of contact with an external reality. This, I take it, is why Ereshefsky and Reydon (2023) connect ICW to their grounded functionality account of natural kinds: While causal reasoning—indeed, *all* reasoning—is tied to the purposes for which we as agents engage in it, in doing so we are tapping into features of the world on which we depend for those purposes to be satisfied. (See Bausman 2023 for an exploration of this "tapping into" notion in terms of adaptation.)

To summarize, ICW begins by supposing that causal reasoning is designed to serve certain purposes and then develops its account in virtue of what purposes it serves and how. First, it holds that we distinguish causal from noncausal relationships because only the former are affordances for control and that causal models are designed to reveal these affordances in a general-purpose format. Second, causal information is expressed—can only be expressed—as information about how a system would change under various interventions from outside it. Third, the interventions contained in the meaning of causal claims are idealized in various ways in order to lend objectivity to our causal models; we aim to make claims that are true or false independently of our beliefs about a system or our ability to intervene on it. These hypothetical interventions are impossible to perform; nevertheless, in forming and testing causal hypotheses, we aim where possible to *approximate* those ideal interventions through controlled experiment.

3. INFORMATIONAL REASONING: COORDINATION AND OBSERVATION

The previous section sketches the key features of Woodward's interventionist framework for causality that I will adapt in developing a general account of information of the kind represented by the use-consensus. A key lesson of ICW, as I understand it, is that the concept of causality is *pre-theoretic*: Calling a relationship causal per se does not commit to some specific theory, from physics or somewhere else, of what grounds or underwrites that relationship. Instead, what we mean when we make such a claim is that it is a potential "lever" for bringing about changes. Of course, much of scientific inquiry involves figuring out what kind of levers they are—for example, by developing mechanistic models for how that phenomenon is realized. However, its being causal *qua causal* is independent of these specific details. What's more, focusing in on these details does not reduce away the "intervention" aspect: Firstly, because intervention remains essential for distinguishing the relevant from the irrelevant properties in those models with respect to the phenomenon in question. Secondly, because abstraction is a vital aspect of explanatory generalization even when the more concrete details are available (Levy and Bechtel 2013).

In short, I take causality in ICW to be more than just a "thin concept" (Cartwright 2005)—that is, more than a placeholder for a range of richer, more informative, context-dependent "thick" concepts. Instead, I take causality in the interventionist sense to be an indispensable abstraction: it is the feature that those relations all share despite their differences—a feature that is vitally important to agents who interact purposefully with the world around them.

This idea will also be an overarching motivation for the complementary account of *information* that I develop here. There is already precedent in the literature for the analogous approach to "information"—that it need not have a predefined theoretical underpinning in order to be useful for empirical inquiry. This view is held by Beckett Sterner (2014), for example: following Bergstrom and Rosvall (2009), Sterner argues that it would be antithetical to biological inquiry to define "information" in a way that establishes its extension based on a priori theoretical principles about semantics. Instead, the concept of information serves simply as a diagnostic tool: it is understood in a way that guides and constrains how we gather and organize empirical cases. Given this, the only conceptual assumptions we should bring into our empirical inquiry are those that tell us how to recognize informational processes when we see them. Crucially, those processes once recognized can and should reciprocally update our theoretical notions about what biological information amounts to.

In a similar vein, Kelle Dhein (2020) details a fascinating case study illustrating how and why ethologists ascribe semantic content to behavior. His case focuses on the study of navigation by ants of the genus *Cataglyphis*, who are surprisingly adept at finding the shortest possible route back to their nest despite long and meandering outward journeys. A research program built around these insects has been highly successful at explaining the ants' navigation ability: as it turns out, ants achieve this through a process of "path integration"—by storing a vector representing the distance and direction from the nest that is constantly updated during its journey. It is important to note, however, that the scientists did not begin this process with a particular theory about semantic information and how it manifests in neurophysiology. Instead, the first step is simply the identification of a surprisingly reliable match between an animal's behavior and its circumstances—in this case, the ant's consistent choice of the shortest journey home. Given a surprising observation of this kind, the search for an explanation is guided by the notion that the animal must possess some means of reliably choosing a successful behavior. The search for an explanation for this phenomenon is understood as the search for an adequate information channel exploited by the ant, whatever precise form that channel turns out to take.

I will return to these works in due course. For now, the key point is that, as Sterner and Dhein both illustrate, the use of "information" in these biological contexts does not depend on some precise scientific or philosophical definition of the concept. Instead, it is an operational principle that guides inquiry into a certain kind of biological phenomenon. The phenomenon in question is what I will henceforth call *coordination*—a concept I will elaborate later. For now, I intend this term to capture, as generally as I can muster, any kind of fortuitous or adaptive "match" between a functional entity or process and its circumstances—one that is surprising enough to invite investigation into how that match comes about. "Information" refers to whatever turns out to play a particular role in that process of coordination. A significant feature of the account I develop is that it elaborates on this idea in terms adapted from Woodward's analysis of causal reasoning: informational reasoning pertains to coordination *in the same way that causal reasoning pertains to control*.

One implication of this is that biological information (in the sense shared by the use-consensus) is not fundamentally different from the everyday human uses of "information" that we tend to find less controversial; rather, they are all instances of a general type of reasoning that I characterize here. In other words, information "for us" (human observers) is simply a special case of information "for the organism." To drive this point, I will take a detour from explicitly biological examples and develop this account of informational reasoning, and its relationship with causal reasoning, in a human context. Following that, I will return in section 3.3 to the case of information in biological practice to show how it exemplifies that same form of reasoning.

3.1 Ideal Observations: The Case of Measurement

Consider a familiar physical system of a simple set of mechanical scales, shown in Figure 4.3. This system consists of a variable *load* representing different weights that might be applied to the scale and a variable *readout* representing different values shown by the dial.

$$load \longrightarrow readout$$

Figure 4.3. A DAG representation of a generic scale. By design, the load placed on the scale causally affects the value displayed on the readout.

To simplify, assume that the possible weights take integer values (e.g., *load* = [0 kg, 1 kg, 2 kg, . . .]) and that the output values are similarly discrete (e.g., *readout* = ["0 kg," "1 kg," "2 kg," . . .]). As one would hope, there is a causal relationship between these two variables: under some range of normal background conditions and within a given range of variation, changing the weight on the scale moves the dial on the readout. This causal relationship is what the edge between the variables represents.

Now, consider the obvious point that this particular physical setup is of a special kind: it is a *measuring instrument.* Hence, we can ask the following question: Is what we've said about this system so far sufficient to account for what makes it a measuring instrument? The answer seems to be "no": in general, for some Y to count as a *measurement* of some X, it isn't sufficient that X causes Y. The reason is stated simply by Peter Kosso (1992) (in the context of "observation"): "An observation, *as it is to contribute to knowledge by motivating and testing claims about the world,* must be an epistemic event and not just a physical event. Not only must the object causally affect the viewer, it must produce an *informative effect* of some sort" (26, emphasis added).

In other words, while measurements (and observations) may certainly involve a causal relationship between the input and the output, measurement is more than that: it is essentially about producing *knowledge* about the measured object. A similar sentiment is expressed by van Fraassen (2008): the outcome of a measurement is a *representation,* with all the intentionality that entails and which causal relationships, qua causal, are lacking. This is one reason I begin my analysis of information with a human example: in biological cases, it may be possible in principle to understand biological processes without intentional terms, opting instead for more physicalist language of causal role functions and so on. But this, I think, is a symptom of the fact that in those cases we are viewing both the object and its perceiver from the third person, from which it's easier for us to see both as just physical systems that are causally related in some way. It is harder to justify when we and our measurements *are* one of those systems. Embedded in the here and now of contemporary scientific practice, we cannot avoid taking the results of our measurements—and things derived from them such as data models—to be *about* something (van Fraassen 2008; Finkelstein 1994; Cropley 1998a; 1998b).

What must we add, then, to make sense of this measurement setup *as a measurement setup*? As suggested by Kosso and van Fraassen, we must clarify the notion of a measurement not merely as a causal process but as a conduit for *information*—of the outcome having an "informative effect." With that in mind, an obvious starting point would be to point out that (under certain assumptions) the load's causal influence on the readout implies that the two will share *mutual information* according to the mathematical theory of communication, or *MTC* (Shannon 1948; 1949). This is the case whenever the values of two variables are correlated; in that case, it's said that one variable "reduces uncertainty" about the other, in the sense that knowing the value of one variable allows you to make a more reliable guess at the value of the other.

Yet there is reason to doubt that this fully captures the informative effect we're looking for. To see why, suppose that the instrument is poorly *calibrated*, that it consistently overestimates the load by, say, 1 kilogram. Crucially, the miscalibration is invisible to the MTC measure of mutual information: as long as every value of *load* corresponds to exactly one value of *readout,* mutual information is at a maximum and there appears to be nothing wrong. Yet calibration is a necessary part of good measurement procedures (Soler et al. 2013), and so neither causality per se nor the mutual information that it underwrites is sufficient as a means of evaluating the scale's adequacy as a conduit for information.

Or so it may seem. There is a conflicting intuition: Surely as long as we *knew* that the readout overestimates the load, we could use the instrument without issue simply by subtracting 1 kg from the indicated value after the fact. To point this out is to say that, *in principle*, there is nothing wrong with the scale itself: the information has made it to the readout intact, though we may lack the ability or background knowledge to properly acquire it. In this sense, the problem is instead with the *interpretation* of its output by a user. Yet the fact remains that not all bangs and scrapes are lessons; that is, merely being causally in contact with the world is insufficient for being *informed* about it. To be informative, our measuring instruments must be set up so that proper interpretation is possible.

To summarize, we seem torn between two ways of looking at this situation: On the one hand, there is "objectively" nothing wrong with the measurement setup. On the other hand, this objective state of affairs is importantly beside the point: measuring setups must take into account an actual subject or user and make it possible for that subject to acquire knowledge. If it doesn't, the setup is inadequate.

As I will show, we can resolve these competing intuitions by considering information as essentially tied to its use. Again, this in itself is far from new. My contribution is to show that this use-consensus about information can be neatly expressed by adapting the conceptual tools of ICW. Recall that ICW's conceptual device of ideal interventions resolves a similar tension about causality: namely, that it is essentially tied to the notion of intervention from without yet at the same time an "objective" property of systems independently of any actual interventions. I propose that information of the sort discussed here can be understood in an analogous way: as a concept, it is essentially tied to the agential notion of *observation*, yet in an important sense it is objectively "there" to be observed.

Following ICW as closely as possible, we can express this idea using directed graphs, albeit with some embellishments. Consider again our miscalibrated scale represented by *load → readout*. We've established the sense in which their relationship is causal. Our aim now is to analyze the claim that the readout carries *information* about the load in the sense of being a source of knowledge to a user—the as-yet-undefined relation in Figure 4.4.

load ◄· · · · · · · · · · · · · · · · · ·► readout
                    ?

Figure 4.4. The scale is designed so that the readout carries information about the load. What exactly defines this informational relationship?

As I've argued, this sense is not fully captured by the mutual information measure of MTC, since that measure is blind to issues such as calibration that are essential to the notion of a good measurement setup. So if that will not do, how should we interpret the relation instead? Recall that, despite being miscalibrated, there is a sense in which the scale is adequate as a measuring instrument. The intuition is that, *in principle,* it's possible to learn about the load from the readout, and in that sense there is "objectively" information to be acquired about the load whether or not anyone manages to acquire it. To turn this from an intuition into a theory, we need to consider more explicitly the agent that can be said to be "learning" about the load. We can represent this as a variable *A* ("agent"), as in Figure 4.5.

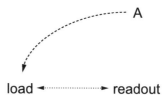

Figure 4.5. An agent *A*'s aim is to learn about—coordinate with—the readout. This coordination relationship is represented by the dashed arrow.

The relationship between *A* and *load* noted previously is what I refer to as *coordination.* We can now clarify this further: *A* refers either to an agent or something used by an agent—for example, a system of representations—whichever is the most useful way of depicting the situation in a given scientific context. We can think of coordination as pertaining to a function $f: A \to load$ that states which value of *A* is appropriate for which value of *load.* One could interpret this as a semantic correspondence relationship in the classical sense—as *load* giving the truth conditions of every value of *A.* However, my aim here is to also include other types of relation that are not (or not obviously) semantic. One non-semantic relation I wish to include is *adaptation:* An adaptive phenotype is not "about" its environment, yet it is *coordinated* with it in the sense I use here. For the sake of characterizing informational reasoning in the abstract, the details of particular instances of this relationship are irrelevant; in any case, the idea that coordination is designed to capture is a match of some kind between the value of *A* and the value of *load.*

A helpful way to understand this idea of coordination is that it relates to ICW's idea of control by inverting its direction of fit. While control is changing the world to suit the agent (or some extension of agency), coordination is fitting the agent to the world. Measurement clearly fits into the latter class: while we may test and calibrate an instrument by manipulating the input to determine its relation to the output, its use *as a measuring instrument* ultimately means pointing the setup at an as-yet-unknown target and hopefully getting a correct value.

With that in mind, suppose that the goal in this case is a coordination between *A* and *load,* and the question is how that might be achieved. The ideal case, of course, would be to have the value of *A* be directly determined by *load.* Yet, as is often the case, we lack such direct access to the object of our interest. In this case, of course, the hope is that the value of *readout* provides this information—that one can instead form true beliefs about the load by conditioning those beliefs on what the readout says. Here we can begin to make this claim both more precise and more general: Whether *readout* contains information about *load,* I propose, is in effect a question of whether *readout* can be exploited to produce a coordination between *A* and *load.*

As we've seen from the previous issue of calibration, there is an apparent tension between two ways of interpreting this question; specifically, about whether to understand this exploitability in relation to actual users, or in a more in-principle sense that is independent of actual users' contingent limitations. Importantly, there is an analogous tension at work in the interventionist notion of causation, and both can be resolved in the same way: recall that causal claims aim to be objective in the sense of not being relative to the contingent knowledge or abilities of actual agents, which is why they are implicitly about *ideal* interventions. I claim that the same thing applies to informational reasoning: In reasoning about the information that is "objectively" in a system, we are in effect reasoning about *ideal observations.* An ideal observation is shown in Figure 4.6.

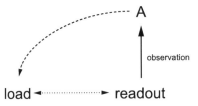

Figure 4.6. By *observation* of the readout, the agent *A* is able to coordinate with the load. This observation is a causal relationship of a particular kind between the readout and the agent.

As the name suggests, ideal observations are information's counterpart to ICW's ideal interventions. Though they share the property of being causal relations, the direction is reversed: while interventions determine the value of their target, observation variables are *determined by* the target of the observation. In other words, while interventions turn their target from a dependent variable in the system to an independent one, observation turns *A* from an independent to a dependent variable. With this in mind, claims about information are about whether and to what extent such ideal observations improve *A*'s chances of coordination; that is, whether hooking *A* up to *readout,* causally speaking, improves the chances of satisfying the coordination function relative to some prior distribution of *A*'s values.

I will avoid committing to certain particulars in how to interpret this question; for example, which interpretation of probability is at work may depend on the particular context to which this general framework is applied. However, there are certain features of ideal observations that should be specified to allow them to do the work required of them; that is, to render the claims about information that they involve "objective" in an important sense.

First, and most obviously, the observation variable must be ideally causally sensitive to its target variable; that is, it should adopt exactly one value for each value of the target. (This allows that more than one value of the observer may be "correct" for the same value of *load*; this is discussed further in section 3.2, but not the reverse.) This is analogous to the condition for ideal interventions that they fully determine the value of their target.

Second, relative to the system in question, the observer variable *A* must be causally independent of every other variable except for the observed one. This requirement is analogous to the condition of ideal interventions that they *uniquely* determine their target and is required for analogous reasons: With interventions, this is to guarantee that any resulting change in the effect is exclusively due to the intervention on the cause. Similarly, ideal observations should guarantee that any resulting coordination with the variable of interest—with *load* in the present case—is exclusively due to the observation of *readout* and for no other reason.[2] For example, suppose we wanted to put palmistry to the test. We may accept that a palm reader can find things out about their subject through a combination of conversation, body-language, "cold reading," and so on. But what we really want to know is whether they can learn anything significant about the person specifically by attending to particular features of their hand. Any controlled test of this claim will have to "block" those other potential sources of information, ensuring that anything the reader can learn about their subject (or at least any *improvement* in their knowledge) is for the claimed reason alone.

Finally, as well as simply being causally sensitive to its target and only to its target, the details of this causal relationship must be of the "right" kind; that is, it must have the right observation function *f: readout → A*. The purpose of this condition is to account for cases of miscalibration as discussed previously: an actual observer may be fully sensitive to the readout, yet assume by default that when the readout says "2 kg," the load is 2 kilograms. An *ideal* observer, however, is by definition one that makes the necessary correction. This captures a notion of error that will be developed more later: the difference between the information the readout *appears* to carry, from an actual user's subjective viewpoint, and what it objectively carries. The idealization simply posits an observer with the necessary background knowledge, in other words, the right observation function.

The overarching purpose of the concept of an ideal observation is to capture the idea of information being an "objective" property of the world: on this framework, to ask whether some *X* carries information about some *Y* is to ask whether it is possible for some agent *with some ideal set of capacities* to coordinate with *Y* by conditioning its state or behavior on *X*. Here lies a critical point: very often, it will simply not be possible for *X* to be used to coordinate with *Y*, no matter what imagined capacities an observer of *X* has. For example, if the scale were broken and the load did not affect the readout at all, there would be no possible way to learn about one by observing the other: changing this fact would mean changing the system itself, not the observer of it. Hence, tying claims about information to idealized observations does not make them trivially true: sometimes information simply doesn't exist to be exploited.

However, that there is information in the world to be exploited in this sense does not mean that actual agents (for example, living things) are able to exploit this information. Nevertheless, ideal observation remains an important concept because it establishes a sense in which there "really is" information to be acquired, independently of whether there are in fact any observers with the right properties to acquire it. Of course, determining what is possible for an ideal observer requires a third-person viewpoint from which we can "see" the whole system—that is, from which we can see both the observed variable and the variable of interest independently. From there, we can determine what kind of relationship (causal, geometric, and so on) exists between the two variables and what capacities an agent would need to have in order to tap into this relationship for some purpose.

Overall, this framework aims to express the sense in which informational reasoning is tied to agency, in just the same way that ICW does. Both causal and informational reasoning are designed to suit the purposes of the agents engaging in that reasoning, hence the presence of interveners and observers in the analysis of the corresponding claims. However, in *idealizing* these interventions and observations, they end up losing their agent-like properties: Intervention and observation variables are simply causal variables with certain particular features. Nevertheless, both causal reasoning and informational reasoning remain intimately tied to agency in that they are designed to reveal affordances—affordances for control in the case of causal reasoning and for coordination in the case of informational reasoning. Yet, as Ismael (2017) argues, causal models reveal these affordances in "generic form": they are not explicitly purposeful because they aim to be neutral about what they might be used for. The same, I claim, applies to information: it is possible to model the ways in which a system *might* conceivably be exploited for coordination without committing to particular purposes for doing so, hence the existence of non-semantic, or *purpose-neutral*, measures of information such as MTC.

A graphical summary of the difference between causation and information is shown in Figure 4.7. This analysis depicts informational reasoning as intimately related to but importantly distinct from causal reasoning: In reasoning about the causal structure of a system, we are reasoning about how it would change if we acted on it in various ways. In contrast, to reason about the information in a system is to reason about what we might learn about one part by observing another. In short, they are two different ways of thinking about a system for different purposes. Given this difference in the basic function of the two reasoning types, each has its own hypothetical, idealized interaction with that system within the types of claim being made. In this sense, then, causation and information are *mutually irreducible.*

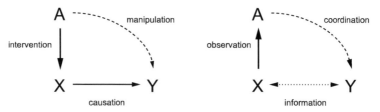

Figure 4.7. The generic form of causal and informational relationships: If *X* causes *Y*, an ideal agent can manipulate *Y* by intervening on *X*. If *X* carries information about *Y*, an ideal agent can coordinate with *Y* by observing *X*.

3.2 What Is the System For—Coordination or Control?

In section 3.3, I'll relate the framework for informational thinking I've developed here to the notion of information in biological practice. Before that, however, I'll offer some further illustration of why informational thinking is in an important sense distinct from causal thinking; that is, why information does not *reduce* to causation in any meaningful sense. This is evidenced by the fact that causal and informational reasoning *evaluate* one and the same physical system in distinctly different ways: a given system may be perfectly adequate as a tool for control and yet *in*adequate as a tool for transmitting information, or vice versa. For one, as is well known, two things don't have to be causally related for one to carry information about the other; correlations without a direct causal link from one to the other are informational nonetheless; for example, they can be related by a common cause. Yet even when the relation in question *is* causal, as with the preceding case of the scale, evaluating that system in terms of the information flowing through that causal process requires a different set of conceptual tools.

To illustrate, consider the causal relationship represented in Figure 4.8. The figure represents not just the causal relationship between X and Y but which *value* of Y results from each possible intervention on X: an intervention setting x_1 produces y_1, while x_2 and x_3 both bring about y_2.

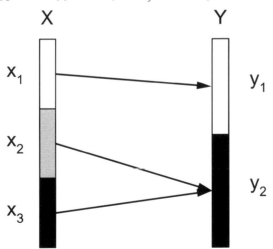

Figure 4.8. A causal relationship between X and Y, showing the possible values of each and how they relate. Here, X taking value x_1 causes Y to take value y_1, while both x_2 and x_3 cause y_2.

First, consider this system from the causal perspective. That is, let us think about this system in terms of its adequacy as a means of *controlling* Y. If this is the aim, then the ideal case would be one in which for any value of Y we might want to bring about, there is an intervention on X that can reliably do so. This is indeed the case in Figure 4.8: there is no value of Y that cannot reliably be brought about by some intervention on X. Hence, provided one *could* intervene on X—remember, whether we actually can doesn't bear on the system's objective features—one could fully control Y. (In fact, if we had such a means at our disposal, we may even come to think of ourselves as setting the value of Y directly!) We can say that x_2 and x_3 are causally *redundant,* and causal redundancy does not threaten control.

Now consider this same system's viability as a means of *transmitting information*—that is, consider the extent to which an *observer* of the effect Y can potentially be informed about (coordinate with) the value of the upstream cause. Again, it is useful to consider the ideal case: a system is an ideal conduit for information when any observed value of Y corresponds to exactly one value of X. (Analogously to what was shown previously, this may lead us to talk as though we have observed X directly.) Importantly, from this perspective the preceding case is *not* optimal: while observing y_1 guarantees that $X = x_1$, observing y_2 only tells the observer—*even an ideal one*—that $X = x_2$ or x_3. To use Dretske's (1981) term, y_2 *equivocates* between x_2 and x_3. The system in Figure 4.8 is therefore adequate as a means of manipulation but inadequate as a conduit for information: information is *lost* between X and Y.

When we consider the reverse case in Figure 4.9, the reverse will be true: In this case, setting x_1 uniquely produces y_1, but setting x_2 can produce either y_2 or y_3. This situation is suboptimal for purposes of control: if we want y_2, there is no way to reliably bring it about through intervention on X. As a conduit for information, however, nothing is wrong: an ideal observer of Y could always coordinate with X perfectly. In doing so, we might interpret the ideal observer as treating y_2 and y_3 as semantically equivalent—or, more generally, by adopting the same state in response to both.

The view I've developed here embodies a kind of dualistic perspective on the relationship between information and the causal properties of a system, though one that should not ruffle a naturalist's feathers. On the one hand, we needn't think of information as a kind of substance that is separate from the physical features of a system: talk of information "in" or "flow-

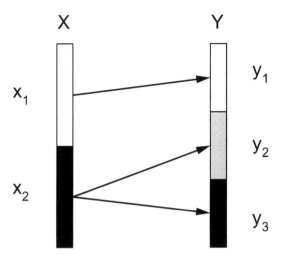

Figure 4.9. In contrast to Figure 4.8, setting X to x_2 can result in Y taking values y_2 or y_3. This situation is also reversed in terms of its adequacy as a conduit for information or control.

ing through" a system can be understood simply as a way in which that system can be used. On the other hand, different kinds of reasoning about a system suit different kinds of use to which that system can be put. So in some sense at least, the two forms of reasoning are designed to access different properties or aspects of that system.

3.3 Coordination Functions in Biology

As I've argued so far, causation and information can be thought of as alternative ways of thinking about a system that each suit different purposes. This makes my account of information and its relationship to causation explicitly *functional*: which perspective is most appropriate to take on a phenomenon depends on what we aim to achieve by doing so. So far I've developed this account in a nonbiological context involving humans and their artifacts. I turn now to how this framework can be used to understand at least one notion of information at work in biological practice. As discussed, a sticking point in the discussion of biological information is in whether we can attribute to information a literal or substantive place in the biological processes we study—whether it is ever information "for the organism," not just for us. Since biology in general is deeply concerned with the purposes or functions of the systems it studies, biological information lends itself well to being understood in functional terms.

As we've seen, the use-consensus broadly takes biological information to be something that is exploited by a biological system in carrying out some function. Sterner and Dhein both put an explicitly pragmatic spin on this idea: Scientists find it useful to *attribute* information to biological phenomena satisfying a certain set of criteria, and doing so serves certain investigative purposes including (but not limited to) guiding investigation that elucidates the mechanisms for those phenomena. In turn, the "information" at work in these phenomena is whatever turns out to play a particular kind of role within those mechanisms. In that sense, information in these practices is not defined by a theory—whether biological, physical, or mathematical. Rather, it is an operational concept that leaves open how particular phenomena are realized and that also permits abstraction and comparison with functionally similar but differently realized phenomena in other species.

I've used the term *coordination* as a label for the kinds of biological phenomena for which this sort of informational reasoning is appropriate. In line with Sterner and Dhein, I hold that (what I call) coordination is presumed by or a precondition of discussions of information in these biological contexts. Dhein's case study of insect navigation is one example: There, investigation begins with the understanding of the coordination the ant is achieving. But biology is replete with others: For instance, the idea of *positional information* (Wolpert 1969) was introduced in the context of explaining how individual cells in a developing embryo differentiate into the "correct" cell type given its position in the mass. Elsewhere, Bechtel's (2009) discussion about the application of *control theory* to cognitive science comes from the idea that brains are functional systems whose sensory capacities serve to detect and respond to environmental circumstances that matter to the organism. What these cases and others have in common is that there is a range of possible states—cell types, behaviors, and so on—that are each appropriate under different circumstances relative to some goal or function. The thing to be explained is how the "correct" choice is reliably made despite the many ways to be *un*coordinated.

However, the reasoning process needn't necessarily *begin* by specifying the coordination relation in question: Sometimes, we begin with a behavior and then ask what state of affairs that behavior is supposed to match with; that is, what its informational function is (Lean 2014). For example, the debate about stotting in gazelles amounts to a discussion of what coordination function it serves: Is it used by other gazelles to indicate danger or by the predator as a sign that the gazelle is too nimble to bother chasing? This is the question whose answer establishes the informational content of stotting as a signal or indeed whether it is a signal at all. So while the study of insect navigation went from an agreed function to the discovery of mechanisms for it, in the case of stotting the "how" was established before the "why." Yet whatever the order, the notion of coordination is a necessary background for the reasoning process. Given some coordination—whether established or simply posited—*information* is whatever makes that coordination possible.

Here we can see why information should not be tied to some particular theory such as MTC: in the case of insect navigation, for example, ants and bees are able to record information about their direction of travel by sensing polarized light. The fact that light polarization can be taken to indicate direction of travel is not readily comprehensible as a statistical relationship between the two, but as something like a *geometric* relation between the light and the ant's orientation.[3] This example illustrates that it is not just statistical regularities in the world that can support reliable inferences and actions; this can also be done by phenomena as diverse as landmarks, one-off causal events, spatial relations, and so on.[4] Yet whatever it is that underwrites the coordination, *what makes that feature of the world "information" is that it is so exploited or that it can be exploited by an ideally endowed observer.* In other words, information can be thought of as an abstract property covering a range of relations in the world that can be exploited to serve coordination functions. Again, this is analogous to my previous claim about causation— that it is an abstract concept linking disparate phenomena by their common capacity to underwrite control.

The third key concept in the triad—observation—is the relationship between the property bearing this informational relationship with the variable of interest on the one hand and the *observer* on the other. My use of "observer" need not imply a strong claim about rich intentionality; we can think of it simply as the thing that is causally responsive to the information carrier, and whose response to that carrier can be seen as bearing a functional match to the circumstances (see Lean 2014). This could apply to part of a mechanism within an organism, such as an *adapter* that links an input to an output for functional purposes (Lean 2019).

In any case, the relationship I call observation is a key component of practices that investigate coordination phenomena in biology. In fact, the *absence* of such a relationship can potentially refute hypotheses about the mechanisms that explain cases of coordination. For example, even if there exists a relationship between light polarization and direction of travel in an insect, this would be irrelevant to the explanation if insects lacked the ability to detect polarized light. That they in fact possess this ability required empirical demonstration of the organism's sensitivity to that feature (Wehner 1997; Rossel and Wehner 1984a; 1984b; Wehner and Müller 2006, all cited by Dhein 2020). Yet by appeal to an *ideal* observer, we can distinguish the information that is objectively "out there" from the subset of that information that is actually playing a role in biological functions. The latter, I suggest, is information in the literal or substantive sense for biological purposes.

Finally, in reconstructing the use-consensus's notion of information in the same way in which ICW understands causation, we have at our disposal a clear sense of how the two are related. This outcome prompts a subtle but important shift in how we think of biological information relative to those who consider it in purely causal terms. In particular, recall the claim by Stotz and Griffiths (2017) that informational relationships in biology are simply causal relationships with high specificity. With the preceding framework in mind, we can interpret the relation between information and causal specificity somewhat differently: rather than saying that informational relationships *are* specific causal relationships, we can instead ask why causal specificity might be useful in systems that are used for informational purposes. Woodward (2010) argues that specific causal relationships are useful in situations when we want fine-grained *control* over some effect. Evidently, this is reasoning in the causal mode—the one that considers the effect of actions. Alternatively, we might instead consider specific causal relationships in the informational mode, in terms of observations and the coordination that they afford. From that perspective, specificity is valuable because it allows fine-grained *discernment* between possible states of the variable of interest, and hence the ability to adjust oneself to those states in a fine-grained way. Similar translations into the informational perspective may be applied to Woodward's other causal properties of stability and proportionality.

In summary, I argued in section 3.1 that while ICW takes causation to be affordances for control, informational reasoning can be understood as reasoning about the affordances a system offers for coordination—about what can be learned, in principle, about one part by observing another. Coordination with one's circumstances through observation—through conditional response—is a basic necessity for agents alongside that of manipulating the world through action. This view lends itself to making sense of when biological systems are trading in information in a literal or substantive sense: this is the case when an organism, or some functional biological system in general, is known or believed to involve some coordination function—to be successfully conditioning its behavior on its circumstances. Relative to that coordination relation, information refers to the relation being exploited by that system to achieve a successful match. In short, biological information exists when something like agency, or at least function, is being attributed to the system. This, I argue, distills in the most general terms the notion of information shared by the use-consensus.

4. CONCLUSION

The methodology by which this chapter relates causation and information aims to embody the lessons of what is sometimes called the "practice turn" (Soler et al. 2014): that is, it aims to consider scientific reasoning as essentially practical, as something we *do* in order to achieve goals in a certain context. This approach is also taken by ICW: it is based on the notion that concepts like "cause" serve certain purposes and can only be fully understood by considering those purposes. In turn, then, the very *origin* of the notion of a cause—the reason we have such a notion in the first place—is because we are actively engaged with the world we're trying to understand. Causality, as we understand it, is intimately connected to *action:* causes are things that can in principle be acted on to influence the world.

I intend my analysis of information in this chapter to be not just analogous but complementary with that view: just as causal reasoning ties to action, informational reasoning ties to *perception:* Information is the world's influence *on us,* insofar as it guides or constrains our inferences and decisions. Just like ICW, however, the aim of objectivity in scientific inquiry leads us to consider information as something that is "out there," in a sense, independently of those inferences and decisions. To meet this need, claims about information can be understood as claims about what a system offers to an *idealized* observer—one that is free of the contingencies and limitations of actual agents. This, I've argued, is complementary to ICW's use of idealized interventions, which serve to express causality as an objective feature independent of actual interveners.

From a naturalist point of view, this conclusion about the complementarity of causal and informational reasoning is unsurprising: Perception and action are the two means by which agents (qua agents) interact with the world around them, and so it makes sense for each to be associated with a distinct type of reasoning engaged in by those agents. Yet those types of reasoning, while in a sense distinct, are intimately related because perception and action are ultimately inseparable: The two jointly constitute an iterative feedback loop, and hence each can only be understood in connection with the other (Dewey 1896; Hurley 2001). Nevertheless it is possible and often useful to decompose that overall process and to view it either from the causal or informational perspective; that is, to view a system either as a target of action or as a conduit for the information that guides action. The fact that biological sciences often treat their objects in at least quasi-agential terms—as doing things for reasons—underwrites what is distinctly biological about biological information.

NOTES

1. Ismael's use of "affordance" is evidently quite different from its use in ecological psychology—a notion I cannot discuss at length here. It should also be noted that Ismael does not subscribe to Woodward's particular interpretation of interventionist causation.

2. However, we can allow a statistical relationship with other variables pre-observation. For example, when one variable screens off another, it may be relevant to point out that the screened-off variable offers no information to some observer over and above what has already been acquired. This may make information claims relative to some prior state of an observer, but it needn't make them subjective.

3. Of course, this evolved system means that light polarization then constrains ants' direction of travel with high probability, which *can* be understood in causal and statistical terms. However, the geometric relationship holds independently of any evolution in the ant's sensory capacities and is the *reason* why those sensory capacities evolved in the first place. Hence, there are two different relations at work here even if the relata are superficially similar.

4. For this reason in particular, the framework I develop should not be taken to lean heavily on the use of directed graphs, which are primarily designed to represent probabilistic relationships. Other means of graphically representing this relationship may be appropriate for other types of coordination phenomena.

REFERENCES

Andersen, H. 2017. "Patterns, Information, and Causation." *Journal of Philosophy* 114, no. 11: 592–622.

Anderson, M. L., and G. Rosenberg. 2008. "Content and Action: The Guidance Theory of Representation." *The Journal of Mind and Behavior* 29: 55–86.

Baggs, E., and A. Chemero. 2018. "Radical Embodiment in Two Directions." *Synthese*. Berlin: Springer.

Barbieri, M. 2016. "What Is Information?" *Philosophical Transactions of the Royal Society A* 374: 20150060.

Bausman, W. C. 2023. "How to Infer Metaphysics from Scientific Practice as a Biologist Might." In *From Biological Practice to Scientific Metaphysics,* edited by W. C. Bausman, J. K. Baxter, and O. M. Lean. Minneapolis: University of Minnesota Press.

Bechtel, W. 2009. "Constructing a Philosophy of Science of Cognitive Science." *Topics in Cognitive Science* 1: 548–69.

Bergstrom, C. T., and M. Rosvall. 2009. "The Transmission Sense of Information." *Biology and Philosophy* 26, no. 2: 159–76. https://doi.org/10.1007/s10539-009-9180-z.

Cartwright, N. 2005. "Causation: One Word, Many Things." *Philosophy of Science* 71, no. 5: 805–20.

Chemero, Anthony. 2003. "Information for Perception and Information Processing." *Minds and Machines* 13: 577–88.

Collier, J. 2008. "Information in Biological Systems." In *Handbook of the Philosophy of Science, Vol. 8: Philosophy of Information*, edited by P. Adriaans and J. van Benthem, 763–87. Amsterdam: Elsevier Science.

Cropley, D. H. 1998a. "Towards Formulating a Semiotic Theory of Measurement Information—Part 1: Fundamental Concepts and Measurement Theory." *Measurement* 24: 237–48.

Cropley, D. H. 1998b. "Towards Formulating a Semiotic Theory of Measurement Information—Part 2: Semiotics and Related Concepts." *Measurement* 24: 249–62.

Dewey, J. 1896. "The Reflex Arc Concept in Psychology." *Psychological Review* 3, no. 4: 357–70.

Dhein, K. 2020. "What Makes Neurophysiology Meaningful? Semantic Content in Insect Navigation Research." *Biology and Philosophy* 35, no. 52.

Dretske, F. I. 1981. *Knowledge and the Flow of Information.* Cambridge, Mass.: MIT Press.

Dretske, F. I. 1988. *Explaining Behavior: Reasons in a World of Causes.* Cambridge, Mass.: MIT Press.

Ereshefsky, M., and T. A. C. Reydon. 2023. "The Grounded Functionality Account of Natural Kinds." In *From Biological Practice to Scientific Metaphysics,* edited by W. C. Bausman, J. K. Baxter, and O. M. Lean. Minneapolis: University of Minnesota Press.

Finkelstein, L. 1994. "Measurement and Instrumentation Science: An Analytical Review." *Measurement* 14, no. 1: 3–14.

Francis, R. 2003. *Why Men Won't Ask for Directions: The Seductions of Sociobiology.* Princeton, N.J.: Princeton University Press.

Garson, J., and D. Papineau. 2019. "Teleosemantics, Selection and Novel Contents." *Biology and Philosophy* 34, no. 36.

Gibson, J. J. 1966. *The Senses Considered as Perceptual Systems.* Boston: Houghton-Mifflin.

Gibson, J. J. 1979. *The Ecological Approach to Visual Perception.* Boston: Houghton-Mifflin.

Griffiths, P. E., et al. 2015. "Measuring Causal Specificity." *Philosophy of Science* 82: 529–55.

Hurley, S. 2001. "Perception and Action: Alternative Views." *Synthese* 129: 3–40.

Ismael, J. 2017. "An Empiricist's Guide to Objective Modality." In *Metaphysics and the Philosophy of Science: New Essays,* edited by M. H. Slater and Z. Yudell. Oxford: Oxford University Press.

Kight, C. R., et al. 2013. "Communication as Information Use: Insights from Statistical Decision Theory." In *Animal Communication Theory: Information and Influence,* edited by U. E. Stegmann. Cambridge: Cambridge University Press.

Kosso, P. 1992. "Observation of the Past." *History and Theory* 31, no. 1: 21–36.

Lean, O. M. 2014. "Getting the Most Out of Shannon Information." *Biology and Philosophy* 29, no. 3: 395–413.

Lean, O. M. 2019. "Chemical Arbitrariness and the Causal Role of Molecular Adapters." *Studies in History and Philosophy of Biological and Biomedical Sciences.* https://doi.org/10.1016/j.shpsc.2019.101180.

Lean, O. M. 2020. "Binding Specificity and Causal Selection in Drug Design." *Philosophy of Science* 87, no. 1: 70–90.

Levy, A., and W. Bechtel. 2013. "Abstraction and the Organization of Mechanisms." *Philosophy of Science* 80, no. 2: 241–61.

Maynard Smith, J. 2000a. "The Concept of Information in Biology." *Philosophy of Science* 67, no. 2: 177–94.

Maynard Smith, J. 2000b. "Reply to Commentaries." *Philosophy of Science* 67, no. 2: 214–18.

Millikan, R. G. 2004. *Varieties of Meaning*. Cambridge, Mass.: MIT Press.

Millikan, R G. 2013. "Natural Information, Intentional Signs and Animal Communication." In *Animal Communication Theory: Information and Influence*, edited by U. E. Stegmann. Cambridge: Cambridge University Press.

Neander, K. 1995. "Misrepresenting and Malfunctioning." *Philosophical Studies* 79: 109–41.

Neander, K. 2017. *A Mark of the Mental: In Defense of Informational Teleosemantics*. Cambridge, Mass.: MIT Press.

Owren, M. J., D. Rendall, and M. J. Ryan. 2010. "Redefining Animal Signaling: Influence versus Information in Communication." *Biology and Philosophy* 25, no. 5: 755–80.

Pearl, J. 2009. *Causality: Models, Reasoning, and Inference*. 2nd ed. Cambridge: Cambridge University Press.

Price, H. 2011. *Naturalism without Mirrors*. Oxford: Oxford University Press.

Rendall, D., M. J. Owren, and M. J. Ryan. 2009. "What Do Animal Signals Mean?" *Animal Behaviour* 78, no. 2: 233–40.

Robinson, A., and C. Southgate. 2010. "A General Definition of Interpretation and Its Application to Origin of Life Research." *Biology and Philosophy* 25, no. 2: 163–81.

Rossel, Samuel, and R. Wehner. 1984a. "Celestial Orientation in Bees: The Use of Spectral Cues." *Journal of Comparative Physiology A* 155, no. 5: 605–13.

Rossel, Samuel, and R. Wehner. 1984b. "How Bees Analyse the Polarization Patterns in the Sky." *Journal of Comparative Physiology A* 154, no. 5: 607–15.

Sarkar, S. 2000. "Information in Genetics and Developmental Biology: Comments on Maynard Smith." *Philosophy of Science* 67, no. 2: 208–13.

Sarkar, S. 2005. "How Genes Encode Information for Phenotypic Traits." In *Molecular Models of Life: Philosophical Papers on Molecular Biology*, 261–83. Cambridge, Mass: MIT Press.

Seyfarth, R. M., et al. 2010. "The Central Importance of Information in Studies of Animal Communication." *Animal Behaviour. Elsevier Ltd* 80, no. 1: 3–8.

Shannon, C. E. 1948. "A Mathematical Theory of Communication." *Bell System Technical Journal* 27, no. 4: 623–56.

Shannon, C. E. 1949. "Communication in the Presence of Noise." *Proceedings of the IRE*, pp. 10–21.

Shea, N. 2007. "Representation in the Genome and in Other Inheritance Systems." *Biology and Philosophy* 22: 313–31. https://doi.org/10.1007/s10539-006-9046-6.

Shea, N. 2013. "Inherited Representations Are Read in Development." *The British Journal for the Philosophy of Science* 64, no. 1: 1–31. https://doi.org/10.1093/bjps/axr050.

Short, T. L. 2007. *Peirce's Theory of Signs*. Cambridge: Cambridge University Press.

Skyrms, B. 2010. *Signals: Evolution, Learning, and Information*. Oxford: Oxford University Press.

Soler, L., et al. 2013. "Calibration: A Conceptual Framework Applied to Scientific Practices Which Investigate Natural Phenomena by Means of Standardized Instruments." *Journal for General Philosophy of Science* 44: 263–317.

Soler, L., S. Zwart, M. Lynch, and V. Israel-Jost. 2014. *Science After the Practice Turn in the Philosophy, History, and Social Studies of Science*. New York: Routledge.

Sterner, B. 2014. "The Practical Value of Biological Information for Research." *Philosophy of Science* 81, no. 2: 175–94.

Stotz, K. 2019. "Biological Information in Developmental and Evolutionary Systems." In *Evolutionary Causation: Biological and Philosophical Reflections*, edited by T. Uller and K. N. Laland, 323–44. Cambridge, Mass.: MIT Press.

Stotz, K., and P. E. Griffiths. 2017. "Biological Information, Causality and Specificity: An Intimate Relationship." In *From Matter to Life: Information and Causality*, edited by S. I. Walker, P. Davies, and G. Ellis, 366–90. Cambridge: Cambridge University Press.

van Fraassen, B. C. 2008. *Scientific Representation*. Oxford: Oxford University Press.

Walker, S. I., et al. 2018. "Exoplanet Biosignatures: Future Directions." *Astrobiology* 18, no. 6: 779–824.

Wehner, Rüdiger. 1997. "The Ant's Celestial Compass System: Spectral and Polarization Channels." In *Orientation and Communication in Arthropods*, edited by Miriam Lehrer, 145–85. Basel: Birkhäuser.

Wehner, Rüdiger, and M. Müller. 2006. "The Significance of Direct Sunlight and Polarized Skylight in the Ant's Celestial System of Navigation." *Proceedings of the National Academy of Sciences* 103, no. 33: 12575–79.

Wolpert, L. 1969. "Positional Information and the Spatial Pattern of Cellular Differentiation." *Journal of Theoretical Biology* 25, no. 1: 1–47.

Woodward, J. 2003. *Making Things Happen: A Theory of Causal Explanation*. Oxford: Oxford University Press.

Woodward, J. 2010. "Causation in Biology: Stability, Specificity, and the Choice of Levels of Explanation." *Biology and Philosophy* 25, no. 3: 287–318.

Woodward, J. 2014. "A Functional Account of Causation; or, a Defense of the Legitimacy of Causal Thinking by Reference to the Only Standard That Matters—Usefulness (as Opposed to Metaphysics or Agreement with Intuitive Judgment)." *Philosophy of Science* 81, no. 5: 691–713.

5

INDIVIDUAL-LEVEL MECHANISMS IN ECOLOGY AND EVOLUTION

MARIE I. KAISER AND ROSE TRAPPES

PHILOSOPHERS HAVE STUDIED MECHANISMS in many fields in biology. The focus has often been on molecular mechanisms in disciplines such as neuroscience, genetics, and molecular biology, with some work on population-level mechanisms in ecology and evolution. We present a novel philosophical case study of individual-level mechanisms, mechanisms in ecology and evolution that concern the interactions between an individual and its environment. The mechanisms we analyze are called niche choice, niche conformance, and niche construction (NC³) mechanisms. Based on a detailed analysis of biologists' research practices, we develop metaphysical claims about the components and organization of NC³ mechanisms, the phenomena they bring about, and how these phenomena relate to individual differences, a major explanatory target in the field. We provide reasons for why processes of niche choice, conformance, and construction are mechanisms and how they differ from molecular mechanisms underlying individual differences. Finally, we demonstrate that a general representation of NC³ mechanisms is highly abstract, such that more specific types of NC³ mechanisms in particular study systems exhibit more complex components organized in more complex ways. Our case study highlights some distinctive features of individual-level mechanisms in ecology and evolution, such as complex and heterogeneous organization and multiple phenomena.

1. INTRODUCTION

How do zebra finch males react to different levels of competition? Why do different buzzards use different kinds of greenery in their nests? What makes a female fire salamander deposit its larvae in a pond or a stream? These are typical sorts of questions asked by behavioral and evolutionary ecologists. To address such topics, behavioral and evolutionary ecologists study individual-level ecological-evolutionary mechanisms. In this chapter, we present a philosophical case study of a paradigmatic example of individual-level mechanisms, so-called niche choice, niche conformance, and niche construction (NC^3) mechanisms.

NC^3 mechanisms reveal how individual organisms interact with their environment, that is, in which activities an individual engages and which acting entities constitute the individual's environment. Specifically, NC^3 mechanisms reveal how individuals change their phenotype–environment match and fitness and, as a result, their individualized niches. Due to their focus on individual organisms rather than populations or sub-organismal entities, NC^3 mechanisms are an instance of "individual-level" (Pâslaru 2018, 349) ecological-evolutionary mechanisms.

The goal of our case study is to contribute to a better understanding of individual-level mechanisms in ecology and evolution. Much of the philosophical work on mechanisms has focused on fields such as cell biology, molecular genetics, and neuroscience (e.g., Machamer, Darden, and Craver 2000; Bechtel 2006; Craver 2007; Craver and Darden 2013). This chapter falls in line with the few philosophical analyses that have been conducted on higher-level biological mechanisms in ecology and evolutionary biology (Baker 2005; Skipper and Millstein 2005; Barros 2008; Pâslaru 2009; 2014; 2018; Raerinne 2011; Havstad 2011; DesAutels 2016; 2018).

Unlike much of this literature, however, our aim is not primarily to consider the extent to which niche choice, conformance, and construction fit into the framework of the New Mechanists (e.g., Machamer, Darden, and Craver 2000; Bechtel 2006; Craver 2007; Craver and Darden 2013; Glennan 2017). Rather, we use this framework to analyze the investigative and explanatory practices of studying NC^3 mechanisms in order to get a deeper understanding of what NC^3 mechanisms are. What makes NC^3 mechanisms special and distinguishes them from, for instance, molecular mechanisms? Which phenomena are explained by describing NC^3 mechanisms, and how do they relate to other kinds of phenomena to be explained in this research field? What components do NC^3 mechanisms have, and how are the components organized?

We begin in section 2 by characterizing our analysis as an instance of "metaphysics of biological practice" (Kaiser 2018b, 29) and by explicating our philosophical methodology. In section 3, we introduce the NC^3 mechanisms, looking at how they involve a focal individual and a focal activity and why biologists refer to them as mechanisms. We then go on to investigate the phenomena that NC^3 mechanisms explain. In section 4, we examine how they explain changes in phenotype–environment match and fitness. Since individual differences are a key topic in the research field that we analyze, in section 5 we consider how individual differences figure in to the NC^3 mechanisms. We argue that NC^3 mechanisms explain a second type of phenomenon, namely changes in individualized niches. Finally, in section 6, we point out how concrete cases of NC^3 mechanisms are far more complex than their initial abstract representation, indicating that simplification is necessary for understanding the commonalities between NC^3 mechanisms.

2. METAPHYSICS OF BIOLOGICAL PRACTICE

Our analysis of NC^3 mechanisms as paradigmatic cases of ecological-evolutionary mechanisms is an instance of "metaphysics of biological practice" (Kaiser 2018b, 29). We develop metaphysical claims about what NC^3 mechanisms are, that is, their ontology. We do so on the basis of careful analysis of scientific practices: how biologists investigate, reason about, and use NC^3 mechanisms to explain certain phenomena. In this section, we specify the metaphysical and the practice-based character of our analysis in turn.

First, we make more than epistemic claims about how biologists represent, study, and explain NC^3 mechanisms. Our analysis involves developing *metaphysical claims* about what NC^3 mechanisms are, which phenomena they bring about, what their components are, how their components are organized, and what distinguishes them from other kinds of biological mechanisms. These are metaphysical (or, more precisely, ontological) claims because they concern what the world is like, which kinds of entities exist in the world, and what these entities are like. Since we draw these metaphysical claims about NC^3 mechanisms from epistemic practices in biology, they can only be "provisional" (Kaiser 2018b, 30). Our metaphysical claims depend on a realistic interpretation of these epistemic practices. That is, we presuppose that the scientific claims made in these practices (or that the practices rely on) are true and that the theoretical terms that they include (e.g., "NC^3 mechanism," "fitness," "environment," or "phenotype") refer to entities that exist in the world independently of scientific investigation. The provisional nature of our claims, however, does not run contrary to their metaphysical character.

Second, we adopt the approach of *practice-based* or "broad-practice-centered" (Waters 2019) philosophy of science. This means that we pay special attention to how biologists investigate NC³ mechanisms, which research questions they pose, how they report about and draw conclusions from their empirical findings, which explanatory strategies they pursue when investigating NC³ mechanisms, and how they individuate these mechanisms and their components.

To do so, we draw on our work as members of a large biological Collaborative Research Centre (CRC) investigating NC³ mechanisms. Being members of the CRC allows us to take into account a broad variety of empirical sources when philosophically analyzing the research practices of the biologists. We analyze the project plans and experimental designs described in the grant application and research talks as well as how the biologists report about the empirical results of their projects in research talks and publications. Where possible, we cite publications, but the ongoing nature of this case study means that many projects are still awaiting final results.

We also gain fruitful insights by directly collaborating with the biologists to refine their central concepts and theoretical assumptions. For instance, we have collaborated with CRC members on a paper for biologists setting out the theoretical framework of the NC³ mechanisms (Trappes et al. 2022). Our approach is thus not only practice-based and "reflective" but also an example of "*embedded* interdisciplinarity" (Kaiser et al. 2014, 66, our emphasis) or what has been called "philosophy-of-science in practice" (Boumans and Leonelli 2013) and "philosophy in science" (Pradeu et al. 2021). Most claims in this chapter, however, result from a reflection about the CRC's research practices and extend beyond our collaborative work with the biologists.

In line with "*empirical* philosophy of science" (Wagenknecht et al. 2015, our emphasis), we also use two qualitative empirical methods to gain more information about the biologists' practices: a questionnaire and interviews. We conducted the questionnaire in October 2018, toward the beginning of the first funding period. Among other topics, we asked members of the CRC open-ended questions about the mechanisms they were researching and how they understood the NC³ mechanisms. There were thirty-seven participants, 90 percent of all scientific CRC members including PhD students, postdocs, and principal investigators (PIs). Responses were analyzed using hypothesis-driven coding, starting with an extensive list of codes as well as grounded and semi-grounded qualitative coding, generating codes while reading responses.

The interviews were conducted in October 2019 with fourteen members of the CRC on their research in the CRC. Interviews were semi-structured with single or paired interviewees. Transcripts were analyzed using hypothesis-driven coding. NC^3 mechanisms were discussed in many of the interviews, both spontaneously and in response to direct questions. In this chapter, we present some results from the questionnaire and interviews and use them in our analysis of the CRC's investigative and explanatory strategies for studying NC^3 mechanisms. For a full description of the empirical methods and further results, see Trappes (2021a; 2021b).

We agree with philosophers such as Ken Waters (2004) that directly asking scientists what they mean by a certain concept is often not the best way to philosophically analyze this concept. At least it should not be the only kind of empirical information that philosophers rely on. Questionnaires and interviews, however, can also be used to analyze what scientists say and how they use a certain concept without directly asking how they understand or define this concept. Waters too acknowledges that empirical methods, such as polls, questionnaires, and interviews, provide a "kind of information [that] could be valuable for the critical analysis of scientific concepts" (2004, 31–32). We think that the philosophical use of empirical methods is particularly fruitful for philosophy *in* science, that is, when philosophers and scientists are members of the same research group and collaborate to pursue the same or closely related research goals. In this case, analyzing the research practices of the scientists by, for instance, examining their research plans and publications, listening to their research talks and discussions, and speaking with scientists is complemented by more structured interactions, such as semi-structured interviews and questionnaires. We therefore use the data gathered from our qualitative empirical methods as one among many resources to develop a broad picture of the biologists' research practices and to generate metaphysical claims about the NC^3 mechanisms.

3. INTRODUCING NC^3 MECHANISMS

3.1 Individual-Level Ecological Mechanisms

Often when we think of mechanisms in biology, we think of things like protein synthesis, gene expression, or neuronal transmission. This is reflected in the literature on mechanisms in philosophy of biology, which has by and large focused on examples of *molecular and genetic mechanisms* in fields such as cell biology, molecular genetics, and neuroscience (e.g., Machamer, Darden, and Craver 2000; Bechtel 2006; Craver 2007; Craver and Darden 2013). This is not to say that biological mechanisms on higher levels and from other biological areas have been ignored. Philosophical work has also been done, for example, on the mechanism of natural selection (Baker 2005; Skipper and Millstein 2005; Barros 2008; Havstad 2011; DesAutels 2016; 2018) and on mechanisms in ecology (Pâslaru 2009; 2014; 2018; Raerinne 2011). Nevertheless, molecular and genetic mechanisms play a dominant role when philosophers think about biological mechanisms.

Interestingly, biologists also tend to associate the concept of mechanism with molecular or genetic mechanisms. In our case study, for instance, the biologists frequently use the term "mechanism" to refer to genetic, epigenetic, transcriptomic, physiological, or hormonal mechanisms that underlie individual differences in behavior or other phenotypic traits. On the other hand, however, they also talk about niche choice, niche conformance, and niche construction as mechanisms. They therefore seem to recognize the existence of mechanisms at higher levels than the molecular.

NC^3 mechanisms can be characterized as "*individual-level mechanisms*" (Pâslaru 2018, 359, our emphasis) because they operate at the level of individual organisms and their abiotic and biotic environment. Usually, descriptions of NC^3 mechanisms identify one (type of)[1] individual as the so-called *focal individual*. It is the individual that takes center stage in the study of an NC^3 mechanism and that engages in the *focal activity* that determines whether the mechanism is one of niche choice, niche conformance, or niche construction. In line with how the concept of an activity is understood in the mechanism debate (Illari and Williamson 2013; Kaiser 2018a), focal activities can be specified as what individual organisms do. They are temporally extended and actualized, and they produce changes (Kaiser 2018a, 120). Besides the focal individual, which is actively involved in a focal activity, there are other entities (passively and actively) involved in a focal activity. In terms of interactions, one can say that all focal activities require that the focal individual interacts with different parts of its abiotic and biotic environment.[2]

In mechanisms of niche choice, conformance, and construction, these individual–environment interactions differ in characteristic ways from each other. As part of our interdisciplinary work in the CRC, we collaborated with the biologists on formulating precise and practically useful definitions of the NC^3 mechanisms (Trappes et al. 2022). According to these definitions, the three focal activities involved in NC³ mechanisms are specified as follows: in niche construction, the focal individual makes changes to its environment; in niche choice, it selects an environment; and in niche conformance, it adjusts its phenotype.

First, in niche construction mechanisms, the focal individual *makes changes to its environment*. This means that the change is happening primarily in the environment rather than in the individual or only in the individual–environment relation. In addition, the individual is actively involved in making these changes (rather than playing a passive role in this activity; Kaiser 2018a, 120). Niche construction mechanisms can involve changes of any abiotic and biotic environmental conditions. This includes altering the presence or abundance of other species, such as when red flour beetles (*Tribolium castaneum*) release quinones, which in turn shapes which microbiota grow in the flour where they live (Project C01; Schulz et al. 2019). Changes to the biotic environment also include changes made to the social environment, such as conspecific behavior or social group size. An example of social niche construction is when harvester ant (*Pogonomyrmex californicus*) queens interact either aggressively or sociably with other queens, which affects the other queens' aggression and sociability and ultimately determines whether they will form a colony together (Project C04; Overson et al. 2014).

Second, the focal individual in niche choice mechanisms *selects an environment* and thereby changes its relation to the environment. Selecting an environment can be understood fairly broadly here, including the individual changing its location, its resource use, or its (social) interactions. For example, in forests in central Germany, female fire salamanders (*Salamandra salamandra*) choose whether to deposit their eggs in free-flowing streams or in standing ponds; this is selection of the reproductive environment, which determines the environment experienced by the offspring (Project A04; Krause and Caspers 2015; Oswald et al. 2020). As another example, researchers tested how cognitive differences affect whether mice (*Mus musculus*) choose to forage in a dangerous environment that has a high reward or in a benign environment that has a low reward (Project A02).

Finally, niche conformance mechanisms involve the focal individual *adjusting its phenotype* in response to certain environmental conditions. Niche conformance therefore involves phenotypic plasticity, the ability to develop different phenotypes, including behavioral, morphological, or physiological traits, in different environments. For instance, one project in the CRC studies how male zebra finches (*Taeniopygia guttata*) alter their levels of aggressive and courtship behavior as well as their ejaculate traits in response to the presence of a male competitor; in this case, the environment to which the focal individual responds is the level of reproductive competition, created by the extra-pair male (Project B04). Another example of a niche conformance mechanism studied in the CRC is the way Antarctic fur seal (*Arctocephalus gazella*) pups (are hypothesized to) change their behavior, hormones, and immune profile in response to the social density of the colony in which they develop; here the environment to which the pups conform is the social density as well as the corresponding parasite load, infection risk, and predation risk (Project A01; Grosser et al. 2019).

In sum, an NC^3 mechanism reveals how a focal individual interacts with its abiotic and biotic environment by either making changes to its environment (niche construction), selecting an environment (niche choice), or adjusting its phenotype (niche conformance). Other authors have grouped these activities together as various kinds of niche construction (Aaby and Ramsey 2019; Chiu 2019). The CRC, however, suggests reserving the term "niche construction" for the more specific mechanism of making changes to the environment, distinguishing as separate mechanisms the other two ways in which individuals interact with their environments (Trappes et al. 2022).

3.2 Why Mechanisms?

As individual-level mechanisms, NC^3 mechanisms are distinct from the lower-level molecular (i.e., genetic, epigenetic, transcriptomic, physiological, or hormonal) mechanisms that underlie individual differences. Given this distinction and the status of molecular mechanisms as paradigmatic biological mechanisms, one might wonder why organism–environment interactions should be described in terms of mechanisms at all. Biologists in the CRC use the term "mechanism" to refer to niche choice, conformance, and construction. This fact alone, however, does not take us very far. In this section, we uncover the biologists' *reasons* for referring to NC^3 as mechanisms. We argue that these reasons are in line with the framework of the New Mechanists (e.g., Machamer, Darden, and Craver 2000; Bechtel 2006; Craver 2007; Craver and Darden 2013, Glennan 2017).

In the questionnaire, we asked participants what makes niche choice, conformance, and construction mechanisms. Various reasons were provided, and the answers were often quite illustrative. For instance, one respondent stated that "all three processes are ways by which individuals can either adjust or adapt to environments. Therefore, they are all tools for an individual to match its phenotype with the environment." Another wrote, "I think of a mechanism very basically as something that describes how a pattern or phenomenon comes to exist/occur. I think all three are potential mechanisms as each describes how individual niches variance can arise within and between organisms."

In total, twenty-three (of thirty-seven) respondents provided reasons for characterizing niche choice, conformance, and construction as mechanisms. A total of twelve respondents stated that NC^3 are mechanisms because they have a specific outcome (five respondents) or because they lead to, result in, or aim at a specific phenomenon (eleven), such as the maximization of fitness, a change in phenotype, or individualized niches. Seven respondents stated that they are mechanisms because they specify how or the way in which a phenomenon is produced (see sections 4 and 5). Two respondents explicitly mentioned explanation in relation to NC^3 mechanisms. In addition, seven respondents mentioned that niche choice, construction, and conformance are processes. One respondent added that they are processes with complex organization, and three respondents emphasized the importance of causal interactions in NC^3 mechanisms. Three other respondents provided reasons that do not match the New Mechanists' framework.

In contrast to the indicative responses from a total of twenty-three of the questionnaire respondents, nine respondents did not provide any reasons for characterizing niche choice, conformance, and construction as mechanisms (six nonresponses, three answers that did not provide any reasons), and five respondents explicitly denied that NC^3 are mechanisms. Four other respondents expressed uncertainty about whether they are mechanisms (in addition to identifying reasons why they might be mechanisms). One explanation for the uncertainty and lack of consensus concerning the status of NC^3 mechanisms is that these concepts were only recently developed within the CRC at the time of the questionnaire. A greater consensus developed during the course of the funding period of the research consortium, as use of the term "NC^3 mechanism" became more commonplace. For instance, in the interviews a year later, nobody questioned the status of the NC^3 mechanisms as mechanisms, and interviewees in nine of the ten interviews talked fluidly about NC^3 mechanisms and what their outcomes are. Another reason for the initial uncertainty and lack of consensus is the difference between underlying molecular mechanisms and NC^3 mechanisms (see section 3.1). Many biologists think of mechanisms in terms of molecular interactions, which was reflected in other parts of the questionnaire (data not reported).

Initial uncertainty notwithstanding, biologists do use the term "mechanism" to describe niche choice, conformance, and construction, and they can generally justify this choice. Most importantly, they refer to NC[3] as mechanisms to emphasize that they lead to *specific outcomes or phenomena* and that describing them explains *how* these phenomena are brought about. This corresponds well with the claims of the New Mechanists that mechanisms bring about specific phenomena (Glennan 1996, 52; Craver 2007, 122), that mechanisms specify how things work (Craver and Darden 2013, 15), and that describing mechanisms thus explains how a specific phenomenon is brought about (Machamer, Darden, and Craver 2000, 2; Bechtel 2006, 27). The references to *causal interactions, processes, and complex organization* are also in line with how mechanisms are often conceived (Machamer, Darden, and Craver 2000; Woodward 2012; Craver and Darden 2013; Glennan 2017).

To conclude, the biologists in the CRC do not use the term "mechanism" arbitrarily, nor is it just a way to lend scientific work an air of credibility. They have plausible reasons to call niche choice, conformance, and construction mechanisms, and these reasons are in line with how philosophers think about mechanisms.

4. EXPLAINING HOW PHENOTYPE—ENVIRONMENT MATCH AND FITNESS CHANGE

4.1 Diverse Explanatory Practices

Standard understandings of mechanisms have it that they bring about specific phenomena; they are mechanisms *of* these phenomena (Glennan 1996, 52; Craver 2007, 122), and describing a mechanism explains the phenomenon that it brings about. Generally, the phenomenon is crucial for the identity of a mechanism, and it determines what is a component of a mechanism and what is not (e.g., Craver and Darden 2013, 52; Kaiser 2018a, 124–26).

What is the phenomenon that NC[3] mechanisms bring about and that the biologists in the CRC seek to explain? When analyzing the explanatory practices in the CRC, we are confronted with a *broad variety of different explanatory strategies and targets,* some of which are very closely related or even overlap. The biologists working on the NC[3] mechanisms investigate how individualized phenotypes and individualized niches arise and change, how a match is produced between an organism's environment and its phenotype, how and why individuals change their environments and phenotypes, how individual differences arise and persist in a population, and how organism–environment interactions affect ecological and evolutionary processes. They do so in concrete study systems that have their own quirks and specificities that demand particular experimental and statistical approaches. Our aim in the next two sections is to show how these various explanatory goals and methods are related and in particular to draw out what it is that the NC^3 mechanisms are supposed to be explaining.

We can first delineate a set of different phenomena that the research projects in the CRC aim to explain. Based on our participatory and empirical work in the CRC, we identified roughly two sets of explanatory targets: changes in phenotype–environment match and fitness on the one hand and individualized phenotypes and individualized niches on the other. This collection of phenomena is evident in the practices of the biologists, as we demonstrate in the following sections. In addition, the existence of various explanatory targets was supported by our empirical work.

In the questionnaire, we asked participants not only what makes niche choice, conformance, and construction mechanisms (the question we looked at in section 3.2) but also what they have in common. The most prominent phenomena cited as the outcome of the NC^3 mechanisms were adaptation or a match between organism and environment (six in each of the two questions about NC^3 mechanisms), an increase in or maximization of fitness (eight for the first question, four for the second question), and individualized niches (five, four). The picture was similar, though slightly different, in the interviews. The most prominent outcome of the NC^3 mechanisms mentioned was individual differences (six of the ten interviews) as well as individualized phenotypes (three) and individualized niches (three). This dominance is likely due to the fact that all interviewees were explicitly asked if NC^3 mechanisms lead to individual differences, individualized niches, and phenotypes. In addition, five of the ten interviews included the idea that NC^3 mechanisms lead to adaptation or a match between an organism's phenotype and its environment, and one interviewee explicitly mentioned an increase in fitness.

In the remaining part of this section, we analyze in more detail the first type of phenomenon, changes in phenotype–environment match and fitness. In section 5, we look at the explanatory targets related to individual differences and identify individualized niches as a second phenomenon explained by NC^3 mechanisms.

4.2 Explaining Changes in Match and Fitness

NC³ mechanisms reveal how focal individuals interact with their abiotic and biotic environment by engaging in one of three focal activities (recall section 3.1). Individuals either make changes to the environment (niche construction), select an environment (niche choice), or adjust their phenotype (niche conformance). What unifies NC³ mechanisms is that all these activities result in the same phenomena. Here we focus on how NC³ mechanisms bring about a change in both the match between an individual's phenotype and environment and the individual's fitness.

On a general level the outcome of NC³ mechanisms and thus the phenomenon that NC³ mechanisms explain can be characterized as the *change in phenotype–environment match and in the individual's fitness.* This phenomenon is brought about by different individual–environment interactions, which can be categorized into the three focal activities introduced in section 3.1. For instance, choosing a different environment (niche choice) will change how well (or badly) an individual's phenotype matches the environment and will change the individual's fitness. Similarly, an individual making changes to its environment (niche construction) or adjusting some of its phenotypic traits (niche conformance) will lead to a change in the phenotype–environment match and in the fitness of the individual.

The fitness concept in the CRC is understood as referring to the number of surviving offspring during an individual's lifetime. Fitness is either measured directly through counting surviving offspring or measured by fitness proxies, such as growth rate, size, time to maturity, or body condition. We are aware of the vast philosophical and biological literature that discusses different fitness concepts and their adequate definitions (for an overview see, for example, Rosenberg and Bouchard 2021). For the purposes of this chapter, however, we do not discuss the fitness concept in any substantial way, but rather adopt the fitness concept of the biologists in the CRC.

The term "match" refers to some kind of intuitive suitability or fitting between an individual (and its phenotype) to the environment that the individual experiences. On an abstract level, we say that a square peg fits into a square hole. Similarly, the skin color of a wild boar matches the vegetation color in German forests (for the aim of camouflage), and the shape of the beak of an avocet matches the conditions in the littoral of the tideland (for the aim of finding food). Phenotype–environment match is thus similar to what is called "ecological fitness" (Rosenberg and Bouchard 2021, sec. 2).

Fitness understood in terms of number of offspring is distinct from phenotype–environment match. However, fitness may be a way of operationalizing match since the increase or decrease of fitness provides the researcher with a way to empirically distinguish matches from mismatches (and better matches from worse). In line with this, fitness is considered by some biologists we work with to be the "ultimate currency" of match. Alternatively, match can also be measured via, for instance, measures of stress levels or well-being, as is common in animal welfare research (Richter and Hintze 2019).

We subsume changes in phenotype–environment match and changes in fitness under one phenomenon because they are closely related in the biologists' practices. In particular, many of the researchers develop hypotheses about how a certain NC[3] mechanism will change both match and fitness. They also design experiments to test these hypotheses, usually measuring phenotypic or environmental outcomes as well as fitness outcomes. For example, biologists study the mechanism of how male zebra finches (*T. guttata*) change their behavior and ejaculate traits in response to the presence of competitors in order to secure more in-pair and extra-pair copulations and fertilization events and thereby increase their chances of having higher reproductive success. The biologists test whether this niche conformance mechanism takes place by looking at whether and how the phenotypic traits change in response to the different environment (with or without competition) and also at how the fitness of the birds (measured as numbers of fertilized eggs and offspring survival) changes as a result. Another example is the niche construction mechanism in which red flour beetles release quinones into their flour and thereby change the microbiota to which they are exposed. The biologists hypothesize that the quinones limit the growth of harmful bacteria and fungi and that more quinones are released when an individual in the group is immunocompromised (for instance, because it has been exposed to a pathogen). Hence, the quinone release is seen as part of a mechanism that might enhance the match between the beetles' phenotypes (for example, whether they are immunocompromised) and their environment (the kinds of flour microbiota). This enhanced match should improve the beetles' fitness by increasing their survival and reproduction. So studying this niche conformance mechanism involves looking at changes in the phenotype–environment match and in fitness.

It is important to note that we characterize the primary phenomenon of the NC[3] mechanisms as a *change,* not an improvement, in match and fitness. Many of the biologists in the CRC hypothesize that NC[3] mechanisms are adaptive, meaning they generate an improvement in match and fitness. Although this is a prominent expectation, it is still posited as a hypothesis to be investigated in concrete cases of each mechanism. In addition, it is assumed that NC³ mechanisms can consist also of nonadaptive evolutionary processes and that some NC³ mechanisms might lead to a decrease in match and fitness (Trappes et al. 2022). We therefore take the primary phenomenon that is explained by NC³ mechanisms to be a change in match and fitness, with the option that it might turn out empirically that these changes are often or generally positive.

The phenomena of NC³ mechanisms can be depicted as in other standard mechanistic diagrams (Craver 2007, 7) as produced by a combination of entities and activities. In Figure 5.1, we develop an abstract depiction of all three NC³ mechanisms, showing how they lead to changes in phenotype–environment match and fitness. In later sections, we will explore and add to

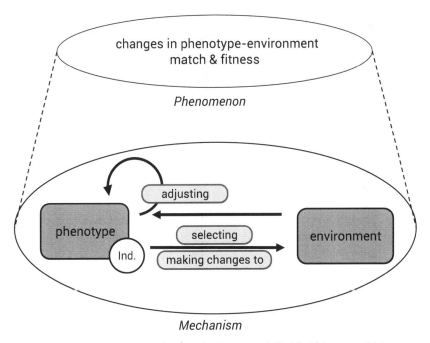

Figure 5.1. Abstract representation of NC³ mechanisms. An individual (Ind.) interacts with its environment, either making changes to or selecting its environment on the basis of its phenotype or adjusting its phenotype in response to the environment. These interactions among the components generate the primary phenomenon of a change in phenotype–environment match and fitness.

this diagram (section 5) as well as introducing diagrams depicting more specific types of NC³ mechanisms (section 6).

One might doubt that NC³ mechanisms are examples of constitutive mechanisms, as the dashed lines in Figure 5.1 suggest. Changes in the fitness of an individual seem to be located on the same level of organization as individual–environment interactions, suggesting that NC³ mechanisms are etiological rather than constitutive mechanisms (on the difference, see Kaiser and Krickel 2017, 751–52). Nevertheless, the change in the match between an individual's phenotype and its environment can be characterized as a process in which the system composed of individual and environment is involved. Consequently, the individual and its environment will be parts of the system involved in the phenomenon and NC³ mechanisms will be constitutive mechanisms.

5. INDIVIDUAL DIFFERENCES AND NC³ MECHANISMS

In section 4.1, we mentioned other phenomena that the CRC seeks to explain. A central goal of the CRC is to understand how individuals differ from one another in their phenotypes and ecological interactions. How does the phenomenon of changes in match and fitness relate to these other phenomena? Our main claim in this section is that explaining changes in match and fitness by describing NC³ mechanisms is interwoven with explaining individual differences in various ways.

The CRC investigates two sorts of individual differences: individualized phenotypes and individualized niches. *Individualized phenotypes* are those phenotypes that differ within a population (and cannot be attributed to obvious population subgroups such as sexes, age classes, and morphs) and are usually stable over time. Paradigmatic examples of individualized phenotypes are color patterns and animal personality (Kaiser and Müller 2021).

Individualized niches, in turn, are ecological niches of individuals that are not shared by all members of a population (nor again by members of sexes, age classes, and morph groups). Individualized niches are modeled on Hutchinsonian ecological niches, multidimensional spaces representing the abiotic and biotic conditions under which a species or population can or does persist indefinitely (Hutchinson 1957; Holt 2009). Since individuals do not persist indefinitely, individualized niches are defined as the conditions under which an individual can survive and reproduce (Takola and Schielzeth 2022; Trappes et al. 2022). Areas within the individualized niche are further distinguished based on a fitness gradient across the various niche dimensions, indicating conditions under which an individual does better or worse. Fitness is thus crucial to defining individualized niches. Individualized niches highlight the multitude of ecological factors and conditions to which individuals relate. However, empirical studies usually focus on particular dimensions of individualized niches, such as prey size, parasite load, or social group size.

Although they are both explanatory targets of the CRC, these two types of individual differences figure in the NC^3 mechanisms in distinct ways. Based on theoretical work with the CRC members as well as an analysis of the CRC's research questions, hypotheses, and experimental designs, we identify the difference as one between components and phenomena. Whereas individualized phenotypes are components of NC^3 mechanisms, individualized niches are a phenomenon brought about by NC^3 mechanisms. We argue for these claims in the next two subsections.

5.1 The Role of Individualized Phenotypes in NC^3 Mechanisms

Recall that the CRC aims to understand how individuals conform to, choose, or construct their environment and why individuals do so in different ways. Individualized phenotypes enter this picture in two ways, depending on which of the three NC^3 mechanisms are at stake.

On the one hand, niche choice and niche construction mechanisms often start with individualized phenotypes. They can thus reveal how individual differences in phenotypic traits, such as behavioral traits, affect how the focal individual brings about changes in the environment (including the social environment). The focal individual's individualized phenotype informs how the individual either makes changes to the environment (niche construction) or selects an environment (niche choice). Hence, one of the central questions addressed by describing niche construction and niche choice mechanisms is: How does the environment change (differently) due to different individualized phenotypes?

Project C04, for example, investigates how individual differences in aggressive behavior determine whether harvester ant queens found a colony alone or cooperatively with other queens. Figure 5.2 represents the main types of components of mechanisms of niche choice and construction and highlights that the causal relation in these mechanisms runs from the phenotype to the environment.[3] An extensive list of examples of projects studying niche construction and choice is included in Table 1 in the Appendix.

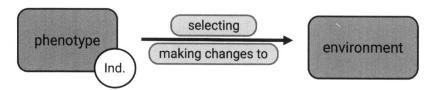

Figure 5.2. Components of niche choice and niche construction mechanisms. An individual (Ind.) selects or makes changes to its environment based on its (individualized) phenotype.

By contrast, mechanisms of niche conformance often start with different environmental conditions or with a change in the environment (including the social environment). They reveal how individuals adjust their phenotypes in response to different or changed environmental conditions.[4] Descriptions of niche conformance mechanisms therefore answer the central question: How do individualized phenotypes change (differently) due to different environmental conditions? In mechanisms of niche conformance, the causal relation thus runs in the opposite direction, from the environment to the phenotype (see Figure 5.3).[5]

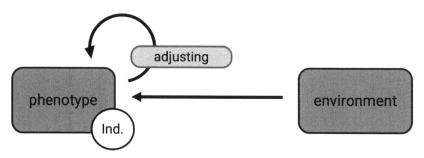

Figure 5.3. Components of the niche conformance mechanism. An individual (Ind.) adjusts its individualized phenotype in response to the environment it experiences.

Many projects in the CRC study niche conformance to explain how individualized phenotypes arise. For example, Project A02 studies how an enriched developmental environment combined with different genotypes influences the level of optimism and pessimism exhibited by adult mice (Krakenberg et al. 2020). Another example is Project B05, which studies the effect of fruit fly (*Drosophila melanogaster*) larval density on adult reproductive and metabolic phenotype. In both cases, the mechanism of niche conformance is proposed to explain how the different environmental factors experienced by individuals lead to different individualized phenotypes.

For an extensive list of examples of projects studying niche conformance, see Table 2 in the Appendix.

Already by looking at individualized phenotypes we can see that the CRC is interested in how different individuals vary with respect to how they choose, conform to, or construct their environments rather than only how individuals generally undertake these activities. Describing NC^3 mechanisms can help explain individualized phenotypes by showing how they are produced in response to the environment (niche conformance) or how they are crucial to the individuals' success in a new environment (niche choice and construction). In addition, regardless of whether we look at niche conformance, choice, or construction, individualized phenotypes are among the components of the NC^3 mechanisms. Recalling that individualized phenotypes require variation between individuals, we can conclude that there can be *components* of the NC^3 mechanisms that vary between different individual instantiations of the mechanism. Indeed, some NC^3 mechanisms require individual differences in the components. For instance, animal personality may in some cases play an essential role in determining which environment an animal selects. We return to this point about variation in section 6.

5.2 The Role of Individualized Niches in NC^3 Mechanisms

Individualized niches are also a central explanatory target of the CRC's investigation of NC^3 mechanisms. Yet they are connected to these mechanisms in a very different way than individualized phenotypes. In this section, we show that changes in individualized niches are an *outcome* of the working of NC^3 mechanisms, albeit in a different way for each of the mechanisms. Specifically, niche choice, niche construction, and niche conformance all change the focal individual's phenotype–environment match and fitness. Together, these imply a change in the individualized niche. Individualized niches are thus located on the phenomenon level, in close relation with the changes in phenotype–environment match and fitness we discussed in section 4.

Take niche choice and construction, which alter the environment side of the phenotype–environment match. On the one hand, choice and construction can introduce new sorts of environmental factors that weren't present before the individual selected or made changes to its environment. In this case, there may be new dimensions added to the individualized niche. For instance, shifting to a new territory may open up new resources for exploitation, thereby adding entirely new dimensions to the individualized niche.

Alternatively, niche choice and construction may shift the range of values taken by an environmental factor in the individual's environment. This would mean that the individualized niche now includes a different range along a preexisting niche dimension. For instance, the red flour beetles studied in Project C01 perform niche construction by releasing quinones, which alter the microbes in the flour. The beetles thereby shift from a higher to a lower range along niche dimensions to do with infection risk by harmful fungal and bacterial species.

Although niche conformance does not involve making changes to the environment, it too affects phenotype–environment match and fitness. By changing its phenotype, the individual will perform better or worse when exposed to certain ecological factors. In other words, niche conformance can change the fitness gradient for a given niche dimension. For instance, the fur seal pups studied in Project A01 conform to the social density they experience, with the hypothesis being that conformance to high density helps pups survive better at a high social density. This means conformance changes the way social density affects fitness, altering the fitness gradient over niche dimensions related to social density. Niche conformance thus alters the individualized niche, not because there are new environmental conditions for the individual to interact with but rather because the individual interacts with the same environmental conditions in a different way.

NC³ mechanisms therefore bring about changes in individualized niches by altering which dimensions make up the individualized niche, which ranges along those dimensions are included in the niche, or the fitness gradient along different dimensions.[6] They do so because they alter the phenotype–environment match and the individual's fitness. There is therefore an intimate connection between the two phenomena of the NC³ mechanisms: changes in phenotype–environment match and fitness imply changes in individualized niches. We depict this relation between the phenomena and the NC³ mechanisms in Figure 5.4.

The phenomena that mechanisms bring about are standardly understood to be important for individuating the mechanisms and their components and boundaries (Craver 2007, 123, 153; Kaiser 2018a, 124–26). Yet it is not entirely clear whether changes in individualized niches are also important for individuating NC³ mechanisms like the change in phenotype–environment match and fitness. Determining whether a change in the individualized niche has taken place generally involves also looking for changes in phenotype–environment match and fitness. This suggests that the changes in match

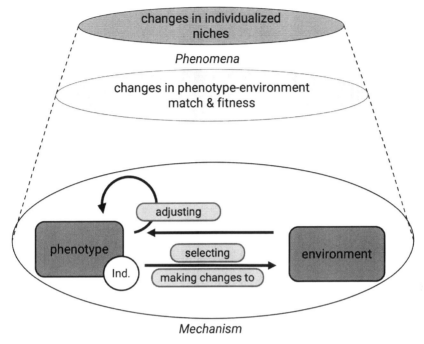

Figure 5.4. Schematic representation of NC³ mechanisms with two phenomena. Interactions between an individual and its environment bring about changes in phenotype–environment match and fitness and thereby alter the individual's individualized niche.

and fitness are used to individuate NC³ mechanisms, with the effect on individualized niches being a derived phenomenon.

Alternatively, changes in match and fitness might gain their interest from their relation to individualized niches. This would mean that the aim to explain changes in individualized niches would be central for individuating NC³ mechanisms, and explaining changes in match and fitness is only an intermediary step to this goal. Understanding and explaining niches is indeed often proclaimed as a target of ecological and evolutionary research, and this transfers to research on individualized niches (Bergmüller and Taborsky 2010; Dall et al. 2012; Trappes et al. 2022). The very names of the NC³ mechanisms highlight the phenomenon of changes to the individualized niche. On the other hand, there is some debate about whether the niche concept is actually the target of explanation, given that researchers usually just investigate a handful of ecological conditions and organism's relations to those conditions (Justus 2019).

It may even be that biologists studying NC^3 mechanisms sometimes focus on one phenomenon and sometimes on the other, in which case the phenomena may deserve equal standing in any general characterization of the NC^3 mechanisms. As with the relation between match and fitness, the role of the two phenomena in the NC^3 mechanisms cannot be decided based on our case study and remains a crucial question for further philosophical research.

6. COMPONENTS OF NC^3 MECHANISMS

We have introduced the focal individual and the three focal activities as the major components of NC^3 mechanisms (section 3.1), specified the two phenomena that NC^3 mechanisms explain (sections 4 and 5), and discussed how NC^3 mechanisms fit into the explanatory practices of the CRC, which focus on individual differences (section 5). Now we can take a closer look at concrete examples of NC^3 mechanisms and analyze how they are studied in the CRC. The goal is to get a more specific understanding of NC^3 mechanisms, in particular what their components are and how these components are organized.

Whereas sections 3, 4, and 5 were concerned with characterizing the general mechanism type "NC^3 mechanism" and its three subtypes "mechanism of niche choice, conformance, and construction," this section analyzes more *concrete examples of NC^3 mechanisms.* These are still types of mechanisms, but much more specific ones (for example, the mechanism of how fire salamanders choose where to deposit their larvae). Even though both parts of our analysis draw on the same research projects and examples, they take place on different levels of abstraction and concern different mechanism types (the more abstract type "NC^3 mechanism" and more concrete types such as "niche choice mechanism in fire salamanders").

Our analysis shows that the general picture presented so far is correct but also *simplistic* in four different ways. First, the focal individual is often not the only individual that is crucial for the working of the mechanism. Second, the focal activity is not as such among the components of NC^3 mechanisms but is rather realized by one or more specific component activities. Third, NC^3 mechanisms involve many more entities and activities than the focal individual, other individuals, and the activities that realize the focal activity. Fourth, individual differences are also not explicit components of concrete NC^3 mechanisms. Together these four claims show that concrete cases of NC^3

mechanisms are much more complex on the component level than suggested by the general picture developed so far and illustrated in Figure 5.4. Looking in more detail at specific cases, we come to better understand the specificities of the components of individual-level ecological mechanisms.

First, recalling section 3.1, NC³ mechanisms are individual-based mechanisms because they reveal how a certain type of individual, the focal individual, interacts with its environment and thereby changes its fitness and how well its phenotype matches the environment. Despite this focus on one focal individual, in most NC³ mechanisms, other conspecific individuals also play an important role. We distinguish two different ways in which *additional individuals* can be involved in NC³ mechanisms.

On the one hand, additional conspecific individuals can constitute the social environment with which the focal individual interacts and that contributes to the change of the focal individual's match and fitness. For example, in the social niche conformance mechanism of zebra finches (Figure 5.5), the focal individual is an adult zebra finch male. The focal male forms a

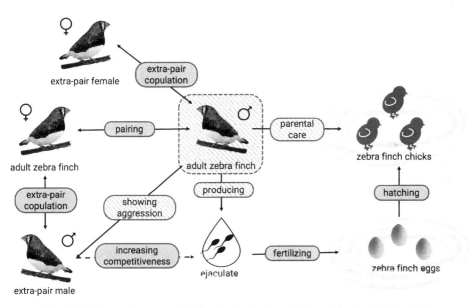

Figure 5.5. Niche conformance mechanism in zebra finches. A focal male (marked with hatching) engages in many different activities in interaction with many other individuals at different times. Three activities—"showing aggression," "producing ejaculate," and "parental care" (marked with hatching)—realize the focal activity of the niche conformance mechanism. The dashed arrow "increasing competitiveness" represents one of the hypothesized effects of an extra-pair male, part of the social environment, on the focal individual's ejaculate traits, an individualized phenotype.

pair with a zebra finch female but also copulates with other females. It also shows aggression toward other males. These other individuals are part of the social environment to which the focal individual conforms, changing its behavior (more or less aggression and parental care) and its ejaculate properties. One of the hypotheses is, for instance, that the presence of an extra-pair male leads the focal individual male to show more aggression and invest less in parental care.

On the other hand, additional individuals can be part of NC³ mechanisms in so far as the focus of the mechanism lies not on one focal individual but rather on a pair or group of focal individuals, which are said to jointly engage in the focal activity. For instance, the buzzard niche construction mechanism (Figure 5.6) involves a pair of adult common buzzards (*Buteo*

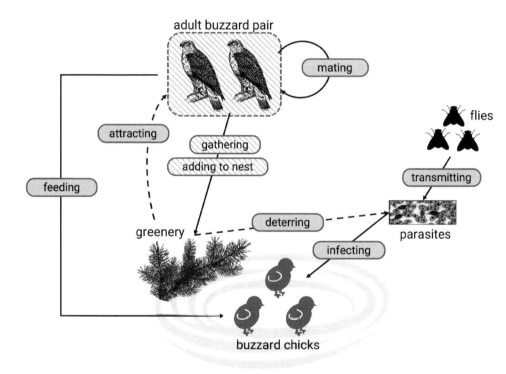

Figure 5.6. Niche construction mechanism in common buzzards. A pair of adult buzzards, both focal individuals (marked with hatching), engage in a number of different activities. The activities that realize the focal activity (marked with hatching) are "gathering" and "adding" greenery to the nest to alter the nest environment. Dashed arrows "attracting" and "deterring" represent hypothesized effects of the greenery on mate quality and parasite levels, respectively.

buteo) jointly engaging in the activities of gathering greenery and adding this greenery to their nest. The biologists hypothesize that the buzzards, by adding greenery to the nest, alter the abiotic and potentially also biotic environment of their offspring. The buzzard pair thereby alters the environmental conditions affecting their own fitness.

Second, when describing concrete cases of NC³ mechanisms, it seems inadequate to refer to the focal activities as such. Activity descriptions such as "making changes to its environment," "selecting an environment," or "adjusting its phenotype" are too general. They need to be specified in concrete cases. Hence, *focal activities* are not among the components of concrete NC³ mechanisms, but rather are *realized* by one or more specific component activities. This can be illustrated in the case of the fire salamander larvae depositing niche choice mechanism (Figure 5.7). The focal activity of selecting an environment, in which the focal individual (the adult fire salamander female) is engaged, is specified by the activity of depositing the larvae in a stream or a pond.

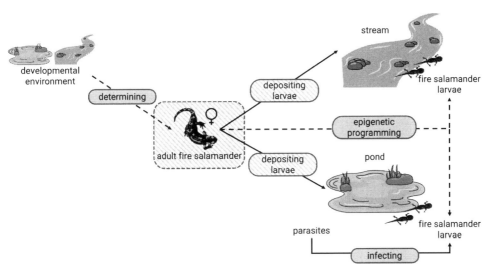

Figure 5.7. Niche choice mechanism in fire salamanders. A focal female (marked with hatching) has a choice between two locations in which to deposit its larvae. The activities that realize the focal activities (marked with hatching) are "depositing larvae" in stream and "depositing larvae" in pond. Dashed arrows "determining" and "epigenetic programming" represent hypothesized ways in which the female's own developmental environment affects its choice and how the female may affect the success of her offspring in the chosen environment.

In several cases of NC³ mechanisms, however, the situation is more complex because the focal activity is realized by a set of different activities and it is not obvious which activities these are. For example, when zebra finch males conform to different social environments (Figure 5.5), the focal activity of adjusting their phenotypes is realized by the zebra finch males showing more or less aggression, engaging more or less in parental care, and producing ejaculate with different properties.

Third, concrete cases of NC³ mechanisms show that the individual–environment interactions can be quite complex and involve *many more entities and activities* than the focal individual, conspecifics, and activities that realize the focal activity. Other entities include parts of the nonsocial environment of the focal individual (for example, the parasites in the ponds infecting the fire salamander larvae, the greenery that the buzzards gather and add to their nests), and other activities include interactions between the focal individual and its social and nonsocial environment that do not realize the focal activity (for example, the extra-pair copulation of the zebra finch males, the adult buzzards feeding their chicks, the developmental environment determining the behavior of the fire salamander females).

What binds together all of these different entities and activities and determines that they are components of a specific NC³ mechanism is that all of them are relevant to the phenomena that the NC³ mechanism brings about (Craver 2007, 123; Kaiser 2018a, 124–26). That is, they all contribute to a certain change in the phenotype–environment match and the fitness of the focal individual as well as a change in the individualized niche (recall sections 4 and 5).

Finally, we discussed earlier (section 5.1) how individual differences in the form of *individualized phenotypes* can be components of NC³ mechanisms. Individualized phenotypes affect how individuals make changes to or select their environments (niche construction and choice), or they result from the adjustment to changed or different environments (niche conformance). As our analysis of three specific cases of NC³ mechanisms shows, individualized phenotypes are not represented explicitly in the NC³ mechanisms. Instead, the mechanisms include general phenotypic traits (such as the behavioral traits showing aggression, depositing larvae, and gathering greenery) without also representing the variation between individuals, that is, how individuals instantiate the traits differently.

In addition, not all traits making up focal activities in the NC³ mechanisms are individualized phenotypes. Some traits may not vary among the individuals in a population, or the variation may not be a focus of the mechanism. For instance, each zebra finch chooses a different mate, and some can be quite picky in making this choice, but mate choice is not a focus of the niche conformance mechanism of how zebra finch males respond to competition.

To conclude, we have seen that the general characterization of NC³ mechanisms in terms of focal individual and focal activity is an abstraction from the details of the actual mechanisms. Other individuals can be involved in various ways, the focal activity often has to be broken down into a number of concrete activities that the focal individual performs, there are other entities and activities involved, and individual differences aren't directly represented. Nevertheless, the more abstract representation helps understand how the very heterogeneous concrete examples of NC³ mechanisms are related, highlighting their similar structure and how they contribute to the target phenomena. This interplay between abstract and concrete, simple and complex, is likely to be a pervasive feature of individual-level mechanisms in ecology and evolution, especially given that causal relations are instantiated in many different ways in different ecological systems (Elliott-Graves 2018).

7. CONCLUSION

Our goal was to generate a better understanding of individual-level ecological and evolutionary mechanisms. To do so, we investigated a paradigmatic case, the NC³ mechanisms. We pointed out how each of the three types of NC³ mechanisms involves a focal individual and a different focal activity. In niche construction, the individual makes changes to the environment; in niche choice, the individual selects an environment; and in niche conformance, the individual adjusts its phenotype. Although they are not molecular mechanisms, we argued that the biologists call niche choice, conformance, and construction mechanisms for plausible reasons that accord with how New Mechanists understand and characterize mechanisms.

We then went on to analyze which phenomena descriptions of NC³ mechanisms explain. The CRC pursues a number of explanatory goals by studying

NC^3 mechanisms. An interesting finding of our case study is that describing NC^3 mechanisms explains more than one phenomenon. NC^3 mechanisms lead to changes in phenotype–environment match and fitness and to changes in the individualized niche. Since the CRC focuses on understanding individual differences, we also clarified how studying NC^3 mechanisms is interwoven with studying individual differences. We found that each sort of individual difference, individualized phenotypes and individualized niches, figured in the NC^3 mechanisms in a different way. Whereas individualized niches are phenomena brought about by NC^3 mechanisms, individualized phenotypes are components of the mechanisms, either a starting point or an end point.

Finally, we discussed the complexities of concrete examples of NC^3 mechanisms. Although we can depict the NC^3 mechanisms in a coherent unified form (Figure 5.4), this representation is highly abstract. It simplifies the multiple individuals, different activities, and other sorts of entities involved in the mechanisms. In addition, it fails to depict individual differences. These complexities, which become evident in more concrete examples of NC^3 mechanisms, can be understood as a consequence of the highly heterogeneous nature of ecological systems. In particular, when ecological systems are considered on the individual level, there are many different individual–environment interactions to be investigated mechanistically.

What can we take away from this case study of ecological-evolutionary mechanisms? Although we do not want to overgeneralize, taking these mechanisms as paradigmatic would open up a number of potential avenues to explore. First, there is the existence of multiple explanatory targets involved in the mechanisms in sometimes complex ways. It seems that mechanisms in ecology and evolution can have multiple phenomena and also that researchers sometimes want to explain changes among the components. Second, there is the interplay between the complexity of concrete cases and the simplification necessary to identify common mechanisms and generate representations of these mechanisms. It seems likely that this feature will recur in other ecological mechanisms, and it would be interesting to investigate further how the more abstract representation of NC^3 mechanisms guides the discovery and description of the more specific NC^3 mechanisms. This would contribute to understanding how organismal biologists employ mechanism schema types as a useful strategy for mechanism construction (Craver and Darden 2013, 71–74).

In addition to expanding our philosophical understanding of mechanisms in ecology and evolution, our work also promises to help biologists gain clarity about the mechanistic nature of individual–environment interactions. In particular, by highlighting the multiple phenomena of NC^3 mechanisms and how individual differences fit in, we provide material for biologists to better justify their own claims about the NC^3 mechanisms. In addition, delineating the structure of focal individual and focal activity, and seeing how they are instantiated in concrete cases by multiple activities and sometimes several individuals, may aid biologists in bridging from their experimental setups and statistical models to the abstract level of NC^3 mechanisms.

ACKNOWLEDGMENTS

This chapter was written in the project D02 "The Ontological Status of Individualised Niches" of the DFG-funded Transregio-Collaborative Research Center "A Novel Synthesis of Individualisation Across Behaviour, Ecology and Evolution: Niche Choice, Niche Conformance, Niche Construction (NC^3)" (SFB TRR-212). We would like to thank all members of the CRC for their insights into their work. Special thanks go to Ulrich Krohs and Behzad Nematipour; many of the ideas in this chapter were developed together through our ongoing discussions. We are also grateful for fruitful comments from the Philosophy of Biology Group at Bielefeld University, from William Bausman and Janella Baxter, and from two anonymous reviewers.

FUNDING

This research was funded by the German Research Foundation (DFG) as part of the SFB TRR 212 (NC^3)–Project number 316099922.

Table 1. Selected projects studying NC³ mechanisms to explain individualized niches. Research questions, explananda, and explanatory factors are adapted from the funding application and research publications from the CRC.

Project (PI)	Research Question	Explanandum individualized niche dimensions	Explanatory factors phenotype, other factors	Mechanism
A01 (Hoffman)	Why do mothers choose different breeding sites?	Social density at breeding site	Fitness trade-offs at each site	Niche choice
A02 (Richter and Sachser)	How does optimism/pessimism affect which niche a mouse chooses?	Risk level of niche	Optimism/pessimism	Niche choice
A03 (Schielzeth)	Do predictability and color affect choice of habitat complexity?	Complexity of habitat	Predictability of escape behavior, color	Niche choice
A04 (Caspers)	How do female salamanders choose their mates?	Assortative mating	Microbiome, chemicals, etc.	Niche choice
A04 (Caspers)	Why do female salamanders choose where to deposit their larvae?	Larval deposition in stream or pond	Developmental programming of phenotype, epigenetics	Niche choice

C01 (Kurtz)	How does individual experience affect the production of quinones the flour microbiota?	Amounts and sorts of microbiota in flour	Exposure to immune challenge, immune status	Niche construction
C03 (Krüger)	Why do buzzards use different greenery in their nests?	Differences in nest greenery, parasite abundance	Parent quality, location of nest	Niche construction
C04 (Gadau)	Do individual differences in aggressive behavior affect colony founding type?	Colony type (mono- or polygynous)	Aggressive behavior	Niche construction and niche choice
D03 (Wittman)	How does niche choice affect the maintenance of spatial variation in density?	Spatial variation in density	Various conditions (specified in models)	Niche choice
D05 (Schielzeth and Reinhold)	Do complex habitats allow more scope for niche specialization via NC³ mechanisms?	Niche specialization (individualized niches generally)	Habitat complexity	Niche choice, conformance and construction

Table 2. Selected projects studying NC³ mechanisms to explain individualized phenotypes. Research questions, explananda, and explanatory factors are adapted from the funding application and research publications from the CRC.

Project (PI)	Research Question	Explanandum individualized phenotypes	Explanatory factors environment, other factors	Mechanism
A01 (Hoffman)	How does social density affect pup phenotype?	Pup behavior, hormones, immunity, transcription, growth, skin microbiome	Social density, maternal programming	Niche conformance
A02 (Richter and Sachser)	How do mice become optimists or pessimists?	Optimism/pessimism	Developmental environment (level of enrichment), genes	Niche conformance
A03 (Schielzeth)	Why do grasshoppers differ in predictability?	Predictability of escape behavior	Habitat complexity, camouflage, individual condition (health)	Niche conformance
A04 (Caspers)	How does larval habitat affect phenotype?	Larval behavior, immunity, skin microbiome, color pattern, etc.	Larval habitat (pond or stream, chosen by mother), maternal programming	Niche conformance
B01 (Kaiser and Sachser)	How do social niche transitions shape biobehavioral profiles?	Endocrinology and behavior	Social group size, social rank	Niche conformance

B02 (Müller)	How does developmental environ- ment affect phenotype?	Larval and adult behavior, physiology, immunity	Developmental environ- ment (access to resources, access to clerodendrins)	Niche conformance
B04 (Korsten and Schmoll)	How is male competitive and parenting behavior affected by sexual competition?	Behavior (competition, mating, parenting), reproductive morphology	Social environment (level of sexual competition)	Niche conformance
B05 (Fricke)	How does population density during development affect adult phenotype?	Adult metabolic & reproductive phenotype	Population density	Niche conformance

NOTES

1. Most empirical studies of NC^3 mechanisms work with groups of individuals that share a particular trait or ecological requirement or relation. Accordingly, most NC^3 mechanisms are represented on the type level, not on the level of token mechanisms with token individuals and activities. For reasons of simplicity, however, we leave out the add-on "type of" in the following.

2. We assume that interactions are activities in which more than one object is involved. A more literal reading of "inter*actions*" would require that the objects that interact have an *active* role. Usually, however, the term "interaction" is used in a wider sense to include activities where only one party is active, a usage we follow.

3. Other factors are also sometimes used to explain the changes in environment, including the environment previously experienced by an individual or the existence of fitness trade-offs for a certain choice. We ignore these for simplicity.

4. Often individuals are confronted with an environment that has changed, meaning researchers study how different individuals conform to the new environment. In other studies, however, individuals face different environments, and the question is how different individuals adjust their phenotypes in response to these different environments. In experiments, these environmental differences are typically restricted to a binary treatment that is intended to represent a continuum. For instance, zebra finches are exposed to two different levels of competition (low or high competition), and drosophila larvae are exposed to different social densities (low or high density). Nevertheless, the researchers acknowledge that in nature the differences are much more gradual (Trappes 2021b).

5. Additional explanatory factors include genotype and maternal epigenetic programming. Again, for simplicity, we leave these out of the representation of the mechanisms.

6. It is also possible to distinguish the effects of the NC^3 mechanisms on *realized* (or actual) individualized niches and *fundamental* (or potential) individualized niches (Trappes et al. 2022). For simplicity, we leave this distinction out.

REFERENCES

Aaby, B. H., and G. Ramsey. 2019. "Three Kinds of Niche Construction." *The British Journal for the Philosophy of Science.* https://doi.org/10.1093/bjps /axz054.

Baker, J. M. 2005. "Adaptive Speciation: The Role of Natural Selection in Mechanisms of Geographic and Non-Geographic Speciation." *Studies in History and Philosophy of Biological and Biomedical Sciences* 36, no. 2: 303–26. https://doi .org/10.1016/j.shpsc.2005.03.005.

Barros, D. B. 2008. "Natural Selection as a Mechanism." *Philosophy of Science* 75 (July): 306–22. https://doi.org/10.1086/593075.

Bechtel, W. 2006. *Discovering Cell Mechanisms: The Creation of Modern Cell Biology.* Cambridge: Cambridge University Press.

Bergmüller, R., and M. Taborsky. 2010. "Animal Personality due to Social Niche Specialisation." *Trends in Ecology & Evolution* 25, no. 9: 504–11. https://doi .org/10.1016/j.tree.2010.06.012.

Boumans, M., and S. Leonelli. 2013. "Introduction: On the Philosophy of Science in Practice." *Journal for General Philosophy of Science* 44, no. 2: 259–61. https://doi.org/10.1007/s10838-013-9232-6.

Chiu, L. 2019. "Decoupling, Commingling, and the Evolutionary Significance of Experiential Niche Construction." In *Evolutionary Causation: Biological and Philosophical Reflections,* edited by K. N. Laland and T. Uller, 299–323. Cambridge, Mass.: MIT Press.

Craver, C. F. 2007. *Explaining the Brain: Mechanisms and the Mosaic Unity of Neuroscience.* Oxford: Clarendon Press.

Craver, C. F., and L. Darden. 2013. *In Search of Mechanisms: Discoveries across the Life Sciences.* Chicago: University of Chicago Press.

Dall, S. R. X., A. M. Bell, D. I. Bolnick, and F. L. W. Ratnieks. 2012. "An Evolutionary Ecology of Individual Differences." *Ecology Letters* 15, no. 10: 1189–98. https://doi.org/10.1111/j.1461-0248.2012.01846.x.

DesAutels, L. 2016. "Natural Selection and Mechanistic Regularity." *Studies in History and Philosophy of Science Part C: Studies in History and Philosophy of Biological and Biomedical Sciences* 57 (June): 13–23. https://doi.org/10.1016/j .shpsc.2016.01.004.

DesAutels, L. 2018. "Mechanisms in Evolutionary Biology." In *The Routledge Handbook of Mechanisms and Mechanical Philosophy,* edited by S. Glennan and P. Illari, 296–307. New York: Routledge. https://doi.org/10.4324/978131 5731544-22.

Elliott-Graves, A. 2018. "Generality and Causal Interdependence in Ecology." *Philosophy of Science* 85, no. 5: 1102–14. https://doi.org/10.1086/699698.

Glennan, S. S. 1996. "Mechanisms and the Nature of Causation." *Erkenntnis* 44: 49–71. https://doi.org/10.1007/BF00172853.

Glennan, S. S. 2017. *The New Mechanical Philosophy.* Oxford: Oxford University Press.

Grosser, S., J. Sauer, A. J. Paijmans, B. A. Caspers, J. Forcada, J. B. W. Wolf, and J. I. Hoffman. 2019. "Fur Seal Microbiota Are Shaped by the Social and Physical Environment, Show Mother–Offspring Similarities and Are Associated with Host Genetic Quality." *Molecular Ecology* 28, no. 9: 2406–22. https://doi.org/10.1111/mec.15070.

Havstad, J. C. 2011. "Problems for Natural Selection as a Mechanism." *Philosophy of Science* 78, no. 3: 512–23. https://doi.org/10.1086/660734.

Holt, R. D. 2009. "Bringing the Hutchinsonian Niche into the 21st Century: Ecological and Evolutionary Perspectives." *Proceedings of the National Academy of Sciences* 106, no. S2: 19659–65. https://doi.org/10.1073/pnas.0905137106.

Hutchinson, G. E. 1957. "Concluding Remarks." *Cold Spring Harbour Symposia on Quantitative Biology* 22: 415–27. https://doi.org/doi:10.1101/SQB.1957.022.01.039.

Illari, P., and J. Williamson. 2013. "In Defence of Activities." *Journal for General Philosophy of Science* 44: 69–83. https://doi.org/10.1007/s10838-013-9217-5.

Justus, J. 2019. "Ecological Theory and the Superfluous Niche." *Philosophical Topics* 47, no. 1: 105–24. https://www.jstor.org/stable/26948094.

Kaiser, M. I. 2018a. "The Components and Boundaries of Mechanisms." In *The Routledge Handbook of Mechanisms and Mechanical Philosophy*, edited by S. Glennan and P. Illari, 116–30. New York: Routledge.

Kaiser, M. I. 2018b. "ENCODE and the Parts of the Human Genome." *Studies in History and Philosophy of Biological and Biomedical Sciences* 72: 28–37.

Kaiser, M. I., and B. Krickel. 2017. "The Metaphysics of Constitutive Mechanistic Phenomena." *The British Journal for the Philosophy of Science* 68, no. 3: 745–79. https://doi.org/10.1093/bjps/axv058.

Kaiser, M. I., M. Kronfeldner, and R. Meunier, 2014. "Interdisciplinarity in Philosophy of Science." *Journal for General Philosophy of Science* 45: 59–70. https://doi.org/10.1007/s10838-014-9269-1.

Kaiser, M. I., and C. Müller. 2021. "What Is an Animal Personality?" *Biology and Philosophy* 36: 1. https://doi.org/10.1007/s10539-020-09776-w.

Krakenberg, V., S. Siestrup, R. Palme, S. Kaiser, N. Sachser, and S. H. Richter. 2020. "Effects of Different Social Experiences on Emotional State in

Mice." *Scientific Reports* 10, no. 1: 15255. https://doi.org/10.1038/s41598-020
-71994-9.

Krause, E. T., and B. A. Caspers. 2015. "The Influence of a Water Current on the
Larval Deposition Pattern of Females of a Diverging Fire Salamander Population (*Salamandra salamandra*)." *Salamandra: German Journal of Herpetol-
ogy* 51, no. 2: 156–60.

Machamer, P., L. Darden, and C. F. Craver. 2000. "Thinking about Mechanisms."
Philosophy of Science 67, no. 1: 1–25. https://doi.org/10.1086/392759.

Oswald, P., B. A. Tunnat, L. G. Hahn, and B. A. Caspers. 2020. "There Is No
Place Like Home: Larval Habitat Type and Size Affect Risk-Taking Behaviour in Fire Salamander Larvae (*Salamandra salamandra*)." *Ethology* 126,
no. 9: 914–21. https://doi.org/10.1111/eth.13070.

Overson, R., J. Gadau, R. M. Clark, S. C. Pratt, and J. H. Fewell. 2014. "Behavioral Transitions with the Evolution of Cooperative Nest Founding by Harvester Ant Queens." *Behavioral Ecology and Sociobiology* 68, no. 1: 21–30.
https://doi.org/10.1007/s00265-013-1618-2.

Pâslaru, V. 2009. "Ecological Explanation between Manipulation and Mechanism
Description." *Philosophy of Science* 76, no. 5: 821–37. https://doi.org/10.1086
/605812.

Pâslaru, V. 2014. "The Mechanistic Approach of the Theory of Island Biogeogra-
phy and Its Current Relevance." *Studies in History and Philosophy of Science
Part C: Studies in History and Philosophy of Biological and Biomedical Sciences* 45 (March): 22–33. https://doi.org/10.1016/j.shpsc.2013.11.011.

Pâslaru, V. 2018. "Mechanisms in Ecology." In *The Routledge Handbook of
Mechanisms and Mechanical Philosophy*, edited by S. Glennan and P. Illari,
348–61. New York: Routledge.

Pradeu, T., M. Lemoine, M. Khelfaoui, and Y. Gingras. 2021. "Philosophy in Science: Can Philosophers of Science Permeate through Science and Produce
Scientific Knowledge?" *The British Journal for the Philosophy of Science*.
https://doi.org/10.1086/715518.

Raerinne, J. 2011. "Causal and Mechanistic Explanations in Ecology." *Acta Biotheoretica* 59: 251–71. https://doi.org/10.1007/s10441-010-9122-9.

Richter, S. H., and S. Hintze. 2019. "From the Individual to the Population—and
Back Again? Emphasising the Role of the Individual in Animal Welfare Science." *Applied Animal Behaviour Science* 212 (March): 1–8.

Rosenberg, A., and F. Bouchard. 2021. "Fitness." In *The Stanford Encyclopedia of
Philosophy*, edited by E. N. Zalta. https://plato.stanford.edu/archives/fall2021
/entries/fitness/.

Schulz, N. K. E., M. P. Sell, K. Ferro, N. Kleinhölting, and J. Kurtz 2019. "Transgenerational Developmental Effects of Immune Priming in the Red Flour Beetle Tribolium castaneum." *Frontiers in Physiology* 10 (February): 98. https://doi.org/10.3389/fphys.2019.00098.

Skipper, R. A., and R. L. Millstein. 2005. "Thinking about Evolutionary Mechanisms: Natural Selection." *Studies in History and Philosophy of Biological and Biomedical Sciences* 36 (June): 327–47. https://doi.org/10.1016/j.shpsc.2005 .03.006.

Takola, E., and H. Schielzeth. 2022. "Hutchinson's Ecological Niche for Individuals." *Biology and Philosophy* 37: 25. https://doi.org/10.1007/s10539-022 -09849-y.

Trappes, R. 2021a. "Individuality in Behavioural Ecology." *OSF*. January 29. https://doi:10.17605/OSF.IO/RKU47.

Trappes, R. 2021b. *Individuality in Behavioural Ecology: Personality, Persistence, and the Perplexing Uniqueness of Biological Individuals.* Doctoral thesis. Bielefeld: Bielefeld University. https://doi.org/10.4119/unibi/2959077.

Trappes, R., B. Nematipour, M. I. Kaiser, U. Krohs, K. van Benthem, U. Ernst, J. Gadau, P. Korsten, J. Kurtz, H. Schielzeth, T. Schmoll, and E. Takola. 2022. "How Individualised Niches Arise: Defining Mechanisms of Niche Choice, Niche Conformance and Niche Construction." *Bioscience* 72 (6): 538–48. https://doi.org/10.1093/biosci/biac023.

Wagenknecht, S., N. J. Nersessian, and H. Andersen. 2015. "Empirical Philosophy of Science: Introducing Qualitative Methods into Philosophy of Science." In *Empirical Philosophy of Science*, edited by S. Wagenknecht, N. J. Nersessian, and H. Andersen, 1–10. Berlin: Springer.

Waters, C. K. 2004. "What Concept Analysis in Philosophy of Science Should Be (and Why Competing Philosophical Analyses of Gene Concepts Cannot Be Tested by Polling Scientists)." *History and Philosophy of the Life Sciences* 26, no. 1: 29–58. https://www.jstor.org/stable/23333379.

Waters, C. K. 2019. "Presidential Address, PSA 2016: An Epistemology of Scientific Practice." *Philosophy of Science* 86 (October): 585–611. https://doi.org /10.1086/704973.

Woodward, J. 2012. "Mechanisms Revisited." *Synthese* 183: 409–27. https://doi .org/10.1007/s11229-011-9870-3.

6
_____

JUST HOW MESSY IS THE WORLD?

JANELLA K. BAXTER

A VIEW THAT IS GAINING IN POPULARITY in the philosophy of
science is that the world is a mess (Waters 2019b; Havstad 2017; McConwell
2017; Dupré 1993; Cartwright 1999). That is, the world that science describes
is characterized by many distinct structures. Philosophers of genetics have
reached this conclusion by arguing that classical genetics and contemporary
molecular genetics are distinct, theoretical, and investigative frameworks
that biologists employ for different purposes (Waters 1994; 2004; 2006;
Weber 2024). What is remarkable is that despite the thoroughgoing pluralism that these authors embrace regarding classical and molecular genetics,
they are nevertheless monistic when it comes to the explanatory and investigative significance of contemporary molecular genetics.

I argue that the pluralism that characterizes molecular genetics is actually more radical than what authors have acknowledged. In fact, the world
of genetics is messier in (at least) two ways. One way has to do with the number and relation of gene concepts at work in contemporary molecular
biology. While Waters and Weber focus primarily on a conception of the
contemporary molecular gene that omits *cis*-regulatory regions, several authors have clarified and defended a number of alternative molecular gene
concepts that treat *cis*-regulatory regions as proper parts (Portin 2009;
Griffiths and Neumann-Held 1999; Stotz 2004; Griffiths and Stotz 2013;
Baetu 2012a; 2012b). I argue further that some genomic databases employ yet
another distinct molecular gene concept—what I call the GenBank gene—
that individuates regulatory sequences as distinct molecular genes on their
own. With a fuller picture of the number of different molecular gene concepts at play in contemporary biology, it becomes apparent that different

gene concepts can overlay and crosscut each other. That is, the same nucleic acid sequence can be classified in a variety of different ways for different purposes. The other way pluralism in contemporary molecular genetics has been mischaracterized has to do with the scope of explanations that appeal to molecular coding genes. I show that the explanatory scope of molecular coding genes can extend beyond the linear sequences of gene products to include observable effects and (sometimes) phenotypic traits.

What my argument shows is that the picture of the world that genetics characterizes is like Cartwright's (1999) idea of a dappled world. The world genetics describes is a patchwork of structures whose boundaries form irregular shapes that can overlay and crosscut each other. Furthermore, the structures that characterize genetics change over scientific history as scientists develop new technologies and practices for managing genomic data. Indeed, the world may be so messy that one might be justified in questioning the usefulness of gene concepts.

1. INTRODUCTION

The world is a mess. Or, at least, this is a position defended by a number of authors in philosophy of science (Waters 2019b; Weber 2023; Havstad 2016; 2017; McConwell 2017; Dupré 1993; Cartwright 1999). This is understood as a metaphysical thesis about the structure of the world. The world that scientists investigate and explain is such that there are a plurality of compatible and distinct frameworks that scientists use to understand it. This is due in part to the way the world is and in part to pragmatic strategies that scientists have for achieving their ends. For example, Joyce Havstad (2016; 2017) has argued that there are multiple justified schemes of protein classification. Proteins have numerous properties and capabilities. Some properties and capabilities are especially useful to scientists for tracking one kind of relation, while others are useful for tracking another kind. For many authors advocating scientific pluralism, pluralism is not a marker of an immature science awaiting replacement by a more mature, fundamental, and unifying theory. Rather, scientific pluralism is "here to stay" (Havstad 2016).

When it comes to the part of the world that genetics describes, just how messy is the world, and what is the nature of the mess? A common view in the philosophy of biology is that there are primarily two distinct, compatible, yet successful frameworks that have characterized the investigative and explanatory pursuits of geneticists since the twentieth century—namely, the classical and molecular (protein and RNA) coding gene frameworks (Waters 1994; 2004; 2006; Weber 2023). The two frameworks are related not by any theoretical or inter-level reduction but instead by a common investigative and explanatory approach. This approach involves using classical or molecular coding genes as tools for producing differences in life processes for the purposes of (primarily) investigating and explaining biological phenomena. The two differ in terms of conceptual structure. The classical gene concept referred to segments of chromosomes whose internal makeup was not known and that are inherited by future progeny according to a set of (relatively) reliable principles and helped scientists explain phenotypic differences in model organism populations. By contrast, the molecular coding gene refers to nucleic acid sequences that encode information about the linear sequences of RNA and proteins and are used to investigate and explain a different set of life processes. Both continue to be successful frameworks in modern biology. Waters and Weber emphasize that the success of these frameworks lies primarily in their experimental purposes. In fact, they maintain that the explanatory scope—or the phenomena explained by the explanans—of these frameworks is quite modest. Gene-centered explanations formulated by Thomas Hunt Morgan and other classical geneticists were often limited to the model organism populations with which they performed experiments. As for the explanatory scope of molecular coding genes, both authors maintain that they only explain the linear sequences of RNA and proteins.

Despite Waters's and Weber's commitment to scientific pluralism when it comes to genetics, they are surprisingly monistic in their attitudes concerning contemporary molecular genetics. In this chapter, I will argue that contemporary molecular genetics is messier than these authors have acknowledged. I do this by defending two theses. The first thesis is that contemporary molecular genetics itself employs a plurality of gene concepts, one of which—the GenBank gene concept—differs importantly from the molecular coding gene concept that focuses on how nucleic acid sequences determine the linear structure of gene products. In defense of this thesis, I draw from major institutional efforts to annotate, curate, and disseminate genome sequence data to communities of scientists. I argue that the classification of *cis*-regulatory gene sequences as genes in the National Center for Biotechnology Information GenBank database has been an effective method for achieving its aims. The second thesis is that the investigative and explanatory scope of molecular coding genes is much more heterogeneous than Waters and Weber claim. I argue that differences in molecular coding genes are often employed in experimental conditions to produce distinctive, observable effects. Not only is this of immense investigative significance, but it is also of explanatory significance. Furthermore, differences in molecular coding genes are also of explanatory significance when it comes to some types of phenotypic traits—like Huntington's disease.

What the arguments of this chapter demonstrate is that, indeed, the world that genetics studies is a mess. But it is messier than what some philosophers of genetics have claimed. The picture of reality that is generated from my arguments is more like Nancy Cartwright's (1999) dappled world notion. The world of genetics consists of more than two major conceptual and investigative structures. It is a world characterized by patches of regularity and structure but whose scope and boundaries are not uniform. The scope and boundaries of these patches are likely to change as scientists develop new techniques and tools to interact with the world. Finally, gene concepts can apply to one and the same sequence of nucleic acid bases. The same sequence can be classified under multiple gene concepts. In light of this image of genetics, it is becoming increasingly hard to say anything more general about the structures that characterize this part of reality.

The structure of this chapter is as follows. In section 2, I outline Waters's and Weber's views about scientific pluralism in genetics. Section 3 demonstrates that their views are not pluralist enough when it comes to the number of gene concepts at play in contemporary molecular biology. In section 4, I argue that their views are not general enough when it comes to the explanatory and investigative scope of the molecular coding gene framework. Section 5 concludes with a discussion of the dappled world of genetics.

2. GENE PLURALISM AND LOCAL EXPLANATIONS

Genes have been the focus of much theorizing and investigation in the life sciences. Yet do not be deceived. The ubiquity of gene-focused science should not be taken to indicate that all mention and use of genes in biology appeals to the same concept. Since the mid-twentieth century, biologists have employed primarily two distinct yet useful gene concepts—the classical and the molecular gene (Waters 2004; 2007; Griffiths and Stotz 2013). For some authors, this has been taken as evidence of scientific pluralism—the thesis that there is no single, fundamental, and comprehensive theory for explaining a given scientific domain of inquiry (Waters 2006; Weber 2023). Waters and (to some extent) Weber have taken this to mean that gene pluralism suggests (at least) two distinct theoretical frameworks for explaining patterns of inheritance and biochemical sequences. Remarkably, notwithstanding their friendly approach to gene pluralism, they have occasionally described contemporary molecular genetics in a surprisingly monistic way.

Since at least the twentieth century, genes have enjoyed a special status in biology. Biologists often conceptualize them as significant difference makers. As the following discussion will reveal, the sense in which they are significant difference makers varies both synchronically and diachronically. That is, at a given point in scientific history (synchronically), there are many ways genes are significant difference makers. Diachronically—or over scientific history—the way genes are significant difference makers can change as well. By "significant difference maker," I mean that a gene or a few genes are singled out as having a causal property that sets them apart from all other relevant causal variables. Genes possess a variety of causal properties that set them apart from all other relevant causes. They can actually (as opposed to potentially) control an outcome of interest, they can determine the fine-grained structure of other biomolecules, they can turn biological processes "on" or "off," and so on. What is relevant for the purposes of this chapter is that the causal property that sets significant difference makers apart from other causal variables is that the property is conceptualized as having a high degree of control over the outcome of interest.[1] Furthermore, the singling out of a significant difference-making variable can also take various forms for different purposes. One way to single out a significant difference-making variable is conceptually. This may be achieved when, say, a model representing the field of all causal variables relevant to some effect of interest is idealized as being fixed or uniform with the exception of a few variables that are imagined as varying. The variables that vary in the model are the significant difference makers. Another way variables are singled out as significant difference makers is experimentally. This is achieved when the field of relevant causal variables are (for the most part) engineered to be fixed or uniform with the exception of the few variables allowed to vary. The two may overlap but needn't always do so. Philosophers of science have done some work to identify the ways in which genes are (and have been) significant difference makers. The purpose of this chapter is to show that contemporary molecular genes are significant difference makers in more ways than have previously been acknowledged. Once the various ways genes are significant difference makers have been parceled out, it will become easier to see what a mess the world of molecular genetics is.

Classical genetics, at least for the first part of the twentieth century in the United States, employed a gene concept that was suited to the investigation and explanation of patterns of inheritance observed in experimental populations.[2] For researchers like Thomas Hunt Morgan, the gene referred to linear units on chromosomes, whose internal structure was unknown (Waters 2004; Kohler 1994). Although the internal structure of the gene was not known, classical genes behave in relatively stable ways according to a handful of principles of inheritance—such as independent assortment, segregation, and recombination (Waters 1994; 2004; 2007). Morgan and his team devised carefully controlled breeding regimens to exploit these principles to generate observable differences in populations of fruit flies that they could attribute to differences in genetics (Bridges and Morgan 1919). Fruit flies with known phenotypic traits, like red eye color, were interbred to generate whole populations that were relatively genetically uniform. This enabled Morgan's group to then breed two distinct genetically uniform populations to generate phenotypic differences. So, for example, a population of red-eye flies might be bred with a population of purple-eye flies. This would generate populations that are heterozygous for the red-eye gene and the purple-eye gene—that is, populations with individuals carrying one copy of the red-eye gene and one copy of the purple-eye gene at the same chromosomal locus. Purple eyes were thought to be a recessive trait, meaning that an individual needs to be homozygous (carrying two copies) for the purple gene to express purple eyes. Another round of interbreeding would generate some individuals who were homozygous for the red gene, some homozygous for the purple gene, and some heterozygous for purple and red genes. Since red eyes were thought to be dominant, only one copy of the red gene is needed for the red eye trait to be expressed. So red eyes were attributed to the presence of red-eye genes, whereas purple eyes were attributed to the presence of two purple-eye genes.

C. Kenneth Waters has emphasized that the success of classical genetics was not its explanatory scope but rather its investigative approach (Waters 2004; 2006; 2010). The explanations Morgan and his team formulated were often modest in scope. When they highlighted the causal significance of classical genes on phenotypic differences, their explanations were partial. Appeals to classical genes only explained the immediate causes of phenotypic differences in experimental populations. Explanations from classical genetics fall incredibly short of all the things one might wish to explain about the origin and maintenance of a trait. Furthermore, the explanations Morgan's group formulated were local. Because the populations used to infer the genetic causes of phenotypic traits were raised and bred in controlled laboratory settings, there were serious questions about how much one is justified in extrapolating findings from Morgan's experiments to wild-type fly populations. Thus (at least on Waters's account), many of the explanations classical geneticists formulated were relative to the experimental populations in which inheritance patterns were observed. If classical genetics was not successful in arriving at highly general explanations—explanations that hold across a wide range of populations, species, genetic and environmental backgrounds, and so on—then why was it successful? Waters's answer: the set of techniques and practices for carefully generating observable patterns in populations whose causes may be reliably inferred is where the success lies. For the purposes of Morgan's group, intervening on genes was an especially effective way to manipulate and control phenotypic traits. Indeed, the gene-centered approach whereby Morgan's group investigated the genotype's causal relationship on phenotype continues to be an indispensable method for much of contemporary biology.

Since the molecular discovery of DNA and the decoding of the genetic code, modern biology has developed another crucial gene concept—the molecular coding gene. Molecular coding genes are sequences of nucleic acid bases—adenine (A), uracil (T), guanine (G), and cytosine (C)—in DNA that encode information for the sequence of a gene product—be it an RNA or protein.[3] Nearby segments of DNA consist of regulatory modules that bind to transcription factors to control the expression of protein coding sequences. Transcription is the process by which a copy of a molecular coding gene (called a messenger RNA or mRNA) is produced according to Watson-Crick base pair rules. When molecular coding genes determine the linear sequence of proteins, they do so during a process called translation whereby mRNA are "read" in units of triplets (or codons) by protein synthesis molecules called ribosomes. Protein synthesis machinery chains together amino acids in a polypeptide according to the (nearly) universal genetic code that associates nearly each codon with one of the twenty canonical amino acids.[4] The genetic code is nearly universal—aside from a few subtle variations, the same triplet associates with the same amino acid across all living species. For example, the UGG codon always "codes" for the amino acid tryptophan and nothing else.

Much like the classical gene, the molecular coding gene also has difference-making capabilities. Molecular coding genes make fine-grained differences to the amino acid sequences of proteins (Waters 2007; Woodward 2010; Weber 2006; 2013; 2017; Griffiths and Stotz 2013). Any given molecular coding gene can have a large number of alternative nucleic acid sequences. Many of the possible alternative sequences a molecular coding gene can take will make a difference to the amino acid sequence of a protein.[5] There is some redundancy in the genetic code, meaning that more than one codon specifies the same amino acid, as in the case of ACU, ACC, ACA, and ACG all of which specify threonine. Furthermore, three codons function as "stop codons"—codons that do not "code" for any amino acid at all but instead carry the information to the protein-synthesis machinery to stop production. Despite these qualifications, a large number of alternative nucleic acid sequences that a molecular coding gene can take will associate with a unique amino acid sequence in a protein. In this way, molecular coding genes can instantiate a large number of differences in the nucleic acid sequences that constitute them, and many of these differences can make many specific differences to the linear sequence of a protein.

The molecular gene concept is flexible and can be used to account for common genomic processes that occur during gene expression. In many eukaryotic organisms, nucleic acid sequences of DNA alone do not always fully determine the linear sequences of gene products (Griffiths and Stotz 2006; 2013; Stotz 2004; Falk 2010). A variety of biochemical processes often have fine-grained causal control over the linear sequences of gene products. An illustrative example of how non-DNA related biomolecules can have fine-grained causal control over the linear sequences of gene products is alternative splicing. In many eukaryotes, the coding sequences (exons) of many molecular coding genes are not continuous but are interrupted by noncoding sequences (introns). During transcription of the gene into mRNAs, introns are cut out and exons are reassembled to form a final mRNA product (called a mature mRNA). Alternative splicing can rearrange, swap, and even scramble the nucleic acid sequence that results in the final mRNA, which in turn determines the amino acid sequence of the protein that is translated. This is how a single protein coding gene sequence can produce more than one—sometimes many—alternative protein sequences.[6] What determines the arrangements of nucleic acids in an alternatively spliced mRNA are cellular environmental conditions. So long as molecular coding genes are identified as the collection of nucleic acid sequences in DNA or RNA that determine the linear sequence of other gene products, the molecular gene concept can accommodate alternative splicing (Waters 2007; Weber 2017).[7] An important consequence of processes like alternative splicing is that the number of molecular coding genes in a genome may be greater than the number of genes annotated by sequencing efforts (Stotz 2006; Burian 2004). Since alternative splicing means that the same annotated gene can give rise to a large number of alternative gene products, a number of molecular coding genes overlap.

As in the case of classical genetics, authors have stressed the modest explanatory significance of molecular coding genes. A common view among philosophers has been that at most molecular protein coding genes explain the linear sequence of other (so-called) information-bearing molecules, like DNA or RNA, and proteins. For example, Waters writes: "The significance of gene-based explanations is modest. Genes can be individuated to explain why particular molecules have the linear structure they do, but that by itself does not explain much" (2019a, 97). Marcel Weber (2023), adopting Waters's view, echoes a similar sentiment:

> As Ken Waters has shown, molecular biology provided a basic theory about how DNA as the genetic material is replicated and expressed. However, unlike so-called "fundamental" theories (as they are thought by some to exist in physics), this basic theory is not able nor does it aspire to explain all the phenomena in its domain. It really only explains how DNA molecules can be copied to produce new DNA molecules with the same or complementary nucleotide sequence (including repair mechanisms), how RNA molecules are processed after transcription, how proteins are synthesized, and how these processes are regulated.

Both authors maintain that although genes encoded in DNA can be appealed to in explanations of extremely limited scope, this is nevertheless a crucial element to the success of contemporary molecular genetics. Again, what is significant about protein coding genes is its use as a tool for intervening and manipulating life processes in an effort not to explain but to investigate the bespoke and complex nature of biological systems (Waters 2010). An important difference between the genetic approach employed by modern biologists and that of classical geneticists is that contemporary biologists have gotten even better at intervening on genes. While classical geneticists had an array of relatively imprecise means of manipulating classical genes—through breeding regimens and crude processes like X-rays—contemporary molecular biologists can now use molecular genes to intervene on other molecular genes. This is best illustrated by the breakthrough gene-editing tool CRISPR-Cas9, whose RNA-guide sequence can be "programmed" to identify almost any nucleic acid sequence in DNA that researchers wish to target and edit with a Cas9 endonuclease (Jinek et al. 2012). Contemporary scientists can determine the precise nucleic acid sequence of a CRISPR-guide sequence, which can in turn aid in introducing precise changes to the nucleic acid sequence of a desired gene in DNA.

Both gene concepts, while related, continue to be useful distinct frameworks.[8] When biologists wish to explain or investigate the inheritance of phenotypic traits that follow regular patterns of independent assortment, segregation, recombination, dominance, and recessiveness, they can employ the classical gene concept. The classical gene concept can be especially useful on the grounds that the exact molecular structure of the gene needn't be known for biologists to utilize it (Griffiths and Stotz 2006; 2013; Waters 1994). By contrast, when biologists wish to explain or manipulate the biochemical processes of life, they can turn to the molecular protein coding gene concept. The older framework is not reducible to the newer one. The two concepts employ different theoretical and practical schemes for explaining and manipulating life processes. What distinct gene concepts show is that there is "no general structure" to genetic explanations and genetic approaches (Waters 2017; also see Griffiths and Stotz 2013). Instead, there is a plurality of frameworks that biologists often switch between fluidly for different purposes.

What is surprising about the authors I have discussed is that despite their thoroughgoing commitment to metaphysical pluralism when it comes to classical and molecular frameworks, their pluralism drops out when they discuss the explanatory and investigative significance of molecular protein coding genes. Of particular interest for my purposes is their exclusion of regulatory regions in DNA from the molecular gene concept. In what follows, I argue that examination of the explanatory and investigative strategies of contemporary molecular biologists unearths an even more radical diversity of structures within the domain of genetics. Not only are there more gene concepts than what has been articulated, but the explanatory scope of molecular genes is more heterogeneous than what has been acknowledged.

3. NOT PLURALIST ENOUGH: THE GENBANK GENE CONCEPT (AND MORE!)

Omitted from the molecular coding gene concept are regulatory regions. *Cis*-regulatory sequences don't have causal control over the linear sequences of gene products but instead control whether and how much molecular coding genes are transcribed. In what follows I trace different ways *cis*-regulatory sequences have been conceptualized in relation to coding sequences by different researchers and for different purposes. In doing so, I show how there have been a plurality of molecular gene concepts at play in contemporary biology. While the determination of linear sequences is an important element to many of the molecular gene concepts I discuss, one concept—notably, the GenBank gene concept—counts *cis*-regulatory sequences as distinct molecular genes. In this way, the GenBank gene concept represents the most notable departure from Waters's molecular coding gene.

Regulatory regions of DNA sequences were (perhaps) first conceptualized as distinct genomic elements by François Jacob and Jacques Monod, whose work in the 1960s characterized the famous lac operon model in prokaryotes (Jacob and Monod 1961a; 1961b; Schaffner 1993; Judson 1996; Keller 2002). The lac operon model is a paradigmatic molecular mechanism consisting of a set of genes (named a, y, z, o, and i) located next to each other in linear fashion along the same chromosomal region (Lac region) (Machamer, Darden, and Craver 2000; Baetu 2012b). In the absence of lactose in the cellular environment, a repressor (encoded by the i gene) binds specifically to the operator (o) gene, which prevents the DNA transcription machinery (RNA polymerase) from producing mRNA transcripts of the z and y genes. The process proceeds until lactose is introduced and binds to the repressor, initiating its release from the operator, thus allowing the transcription of the lactose-metabolizing enzyme β-galactosidase (z) and accompanying genes. Jacob and Monod distinguish between several types of genes, including the operator gene (Jacob and Monad 1961a, 318). The operator gene is a *cis* regulatory region consisting of a nucleic acid sequence embedded in the same strand of DNA as the molecular coding genes whose transcription it controls. Differences in the nucleic acid sequence of the operator gene do not produce differences in the linear sequence of any gene product. Instead, differences in the sequence of the operator gene can alter whether the repressor protein binds to it, thereby turning transcription of the other molecular coding genes in the lac operon "on" or "off" (Jacob 1977; 1988).

Additional regulatory regions were later discovered in other prokaryotes that display different types of control over transcription, such as the L-arabinose operon, which involves a regulatory region with positive control over transcription of the molecular coding genes when bound to a regulatory protein (Englesberg and Wilcox 1974).

In eukaryotes, *cis*-regulatory regions are more diverse and complex than what the operon model suggests. *Cis* and *trans* in this context refer to where a factor lies in relation to the coding sequences with which it interacts. *Cis*-acting regions are located on the same DNA strand as the molecular coding genes they regulate. While the regulatory regions of prokaryotic operons are commonly located nearby and upstream of the molecular coding sequences they regulate, *cis*-regulatory regions in eukaryotes can be found at distal locations either up- or downstream of the molecular coding genes they control (Wittkopp and Kalay 2012). A nucleic acid sequence that is contained within a molecular coding sequence (as either an intron or exon) may, in some contexts, function as a noncoding regulatory region (Gerstein et al. 2007). It is also common for multiple noncoding regulatory regions to be involved in transcriptional control. Different regulatory regions—promoters, enhancers, silencers, insulators, and so on—bind to different types of biomolecules (called transcription factors). Different combinations of transcriptional factors binding to regulatory regions control not only whether a molecular coding sequence is transcribed but the rate and duration of transcription (Griffiths and Stotz 2013).

The Human Genome Project's annotation of regulatory regions raised conceptual questions to how biologists individuate molecular genes. In its guidelines for gene annotation, the Human Genome Project employed a rather inclusive concept of the molecular gene as "a DNA segment that contributes to phenotype/function. In the absence of a demonstrated function a gene may be characterized by sequence, transcription, or homology" (Wain et al. 2002, 464). This conception counts nucleic acid sequences with a wide variety of functional roles—of which the linear sequence-determining capacity of molecular coding genes is only one—as molecular genes. Since differences in noncoding regulatory regions can produce differences in phenotypic traits, *cis*-regulatory regions count as molecular genes on this conception. The Human Genome Project's inclusive gene concept prompted biologists to reevaluate the molecular gene definition (Baetu 2011). Are *cis*-regulatory regions parts of molecular genes—even when distally located from the coding sequence whose transcription they controlled? There is no univocal answer. A great variety of gene concepts have been defended in the philosophical and historical literature; in what follows, I mention only a few.[9] Some scientists advocated for the molecular coding gene concept that forms the heart of Waters's and Weber's concept (Gerstein et al. 2007). Others have argued for individuating molecular genes by the collection of nucleic acid sequences (including noncoding sequences)—however distally related to coding sequences—that determine not only the linear sequence of gene products but the expression as well (Singer and Berg 1991; Piatigorsky 2007).[10] On this conception, noncoding regulatory regions are a proper part of a single molecular gene that includes the nucleic acid sequence that determines the linear sequence of other biomolecules. Yet more radically, some authors have defended what Paul Griffiths and Karola Stotz (2006) have described as the postgenomic gene concept. On this conception, molecular genes are not entities but processes involved in regulating transcription, splicing, editing, and translating coding sequences (Portin 2009; Griffiths and Neumann-Held 1999; Stotz 2004; Griffiths and Stotz 2013; Baetu 2012a; 2012b).

While the various gene concepts discussed thus far concern how scientists think about molecular genes in experimental and theoretical contexts, curators of genome databases play a crucial role in individuating and annotating important genomic elements. Major genome databases—such as GenBank, European Molecular Biology Laboratory (EMBL), and the DNA Databank of Japan (DDBJ)—serve as central hubs for the most updated and comprehensive genome information. Individual research labs deposit nucleic acid sequences to genome databases, where the information is checked for redundancy and accuracy. When novel sequences are deposited, database curators process the data in ways that make it portable across scientific, institutional, and database boundaries (Leonelli 2016). GenBank, EMBL, and DDBJ collect and disseminate genomic information for several crucial purposes. One is to make genome information accessible to facilitate research of labs located all around the world (Benson et al. 2007). Another is to inform the structure and content of other more specialized databases (such as Protein Data Bank and TrEMBL) (Gutierrez-Preciado, Peimbert, and Merino 2009). Managing genomic information requires a delicate balancing act between processing data in ways that are informative but not too informative for users (Kanehisa, Fickett, and Goad 1984; Leonelli 2016). Research labs that use genome databases represent very diverse epistemic communities with different methodological approaches and scientific aims. The success of a genome database depends on how well it makes genomic information accessible to diverse users. Labeling and organizing genomic information is crucial for making entries searchable in a myriad of ways. At the same time, the way curators label and organize data risks biasing future research away from some questions that may deserve attention. Database curators seek to manage this balancing act by employing simple conceptual schemes (Kanehisa, Fickett, and Goad 1984).

Implicit to GenBank's individuation and annotation practices is a simple conceptual scheme that treats *cis*-regulatory regions as distinct molecular genes.[11] On the GenBank gene concept, any nucleic acid sequence that has a confirmed difference-making capacity with respect to some phenotype counts as a distinct molecular gene. As far as the GenBank gene concept is concerned, coding and noncoding sequences are named as "genes." That the GenBank gene concept is distinct from other molecular gene concepts is not lost on maintainers of the database. Instructors running the National Institute of Health's training sessions on how to submit genome sequence data highlight this fact. For example, in Part 1 of "A Submitter's Guide to GenBank," Bonnie L. Maidak states:

> The last set of questions regard the biological meaning of the sequence: and that is what gene does this represent? When I say "gene," sometimes you actually have a sequence not as officially recognized gene, but a genomic region which encompasses a specific genomic marker. You still need to tell us the biological meaning of the sequence even if the sequence that you determine might be an intron of a gene and not the coding region, or it might be just a genomic region but we still need to know why the sequence is important and what the value is of it and what the meaning is of it. ("Webinar" 2014)

GenBank's distinctive gene concept is manifested in the way diverse genome sequences are annotated in the database. For example, the ZPA regulatory sequence (ZRS) has its own entry in GenBank where it has a gene symbol, gene description, gene type, and gene ID number.[12] ZRS is an intron embedded in another molecular coding gene (limb development protein 1 (LMBR1)) that regulates the expression of the sonic hedgehog signaling molecule (SHH) gene (Lettice et al. 2003). Differences in either the nucleic acid sequence or the transcriptional factors that bind to ZRS can associate with significant differences in phenotype like polydactyly in humans (Wu et al. 2016). Of particular concern to the data curators of GenBank are nucleic acid sequences that contribute causally in some way to human health and disease (Sayers et al. 2020). Coding and noncoding sequences alike have this capacity. So, for the purposes of GenBank, a molecular gene is any nucleic acid sequence (coding and noncoding) differences that have a causal effect on a phenotypic trait relevant to biomedicine.

Like other gene concepts, the GenBank gene concept is useful for facilitating investigation and explanation of biological processes. A notable difference between the GenBank gene concept and the molecular coding gene concept is that the former facilitates investigation and explanation of phenotypic differences rather than the linear sequences of gene products. Each entry in the GenBank database provides what Sabina Leonelli (2016) calls a classificatory theory—a concise representation of what the scientific community takes itself to know about the genome sequence rather than novel hypotheses. Each GenBank gene entry includes the full genomic sequence, the sequence's genomic context, gene products (if any), known variants in the population, phylogenetic relationships and organisms carrying orthologues, a brief summary and explanation of the sequence's causal relationship to a biomedically relevant phenotype, alternative names found in the scientific literature, and so on. Also included for each entry is bibliographic information about the sources from which many of the knowledge claims made by the entry come. The GenBank gene concept makes diverse genomic elements searchable within the same database and provides a unifying structure for genomic information that would otherwise be widely dispersed throughout the scientific literature. By making genomic information searchable, the GenBank database helps guide researchers from diverse disciplinary and epistemic cultures to bibliographic and conceptual resources that further their inquiries.

Importantly, the GenBank gene concept—like other gene concepts—is a basic feature of local, partial theories and investigative strategies. Although users of GenBank may explicitly adopt the GenBank gene concept when perusing the database, the concept is implicitly adopted outside the context of database usage to the extent that the GenBank classificatory scheme might organize and direct further inquiry and management of genomic concepts. Furthermore, each GenBank gene will only have causal control over a phenotype for some restricted range of background conditions. Some genes will have causal control over a given phenotype in very restricted genetic and extra-organismal environmental contexts; in other cases, genes will have causal control over a phenotype for a much broader range of background conditions. Even in cases whether the investigative and explanatory scope of a gene is broad, there will nevertheless be much the gene can't explain. For example, differences in the ZRS gene or differences in the transcription factors that bind to may account for polydactyly in humans; however, such differences still won't be able to fully account for all differences in limb differences in humans. What this shows is that explanations featuring GenBank genes have a varied scope.

The world of contemporary molecular genetics is in fact much more pluralistic than what has been suggested by Waters. Molecular genes are in fact individuated by scientists in a variety of ways and for a myriad of purposes. For the purposes of explaining stark differences in a gene product's phenotype, it may be useful to employ a molecular gene concept that incorporates regulatory sequences and processes. The GenBank gene concept serves to signal to diverse scientific communities a common structure shared by incredibly diverse genomic elements for the purpose of directing further research. While GenBank genes differ in terms of the type of causal property that sets them apart from each other, they share the common feature of being heritable genomic units.

A messier kind of pluralism emerges from this picture of gene concepts in contemporary molecular biology than what Waters and Weber have articulated. For one thing, the number of distinct, irreducible concepts is greater than what both authors have indicated. For another, distinct gene concepts don't simply pick out distinct phenomena. That is, contemporary molecular gene concepts can overlay each other. One and the same nucleic acid sequence can be classified under a variety of gene concepts. For example, the limb development protein 1 molecular gene is simultaneously a protein coding gene and a regulatory gene. The same logic Waters employs to reach the metaphysical conclusion that the world is messy and characterized by a plurality of structures should prompt us to accept that the world is in fact much messier than has previously been appreciated.[13]

4. NOT GENERAL ENOUGH: THE EXPLANATORY SCOPE OF PROTEIN CODING GENES

Another way Waters and (in adopting Waters's view) Weber have mischaracterized the significance of gene concepts has to do with the explanatory scope of protein coding genes. Waters and Weber emphasize that protein coding genes are significant difference makers with respect to other biomolecules owing to their fine-grained causal control over the linear sequences of RNA and proteins. I argue that they have understated the explanatory scope of protein coding genes. At least in some cases, protein coding genes can have fine-grained control over phenomena that extend beyond the linear sequences of some biomolecules.

Being a significant difference maker—however "significant" is conceptualized—has explanatory import for Waters and Weber. Difference makers provide answers to *what if things had been different* or counterfactual questions—a property many take to be relevant to explanation. Now, of course, nearly every effect has a great number of difference making variables; however, not all are cited in scientific explanations. Instead, scientists tend to single out some (significant) difference makers as being genuinely explanatory. For this practice to be principled and nonarbitrary, the singled-out difference makers must have some property that sets them apart from all other relevant difference makers.[14] Waters and Weber maintain that the property that sets protein coding genes apart from all other difference makers is fine-grained control over the molecular sequences of other biomolecules (RNA, proteins, etc.). In what follows, I argue the protein coding genes can have this property with respect to phenomena that extend beyond the molecular sequences of biomolecules.

Importantly, Waters and Weber emphasize protein coding genes as tools for manipulating and controlling biological processes. Much of the history of molecular biology has been about making protein coding genes into experimental tools that make observable otherwise unobservable phenomena. Fluorescent proteins are an illustrative example. These proteins are crucial observable technologies employed by biologists in laboratories globally. The molecular protein coding genes for fluorescent proteins used in biology have been modified in various ways. The gene for green fluorescent protein has been modified to have spectral properties that work best with laboratory microscopes (Chalfie et al. 1994). Other genes for fluorescent proteins have been modified to encode proteins that fluoresce new colors (Shen et al. 2017). It is becoming more common for structural biologists to study the three-dimensional structure of proteins by manipulating the nucleic acid sequences of protein coding genes to alter the amino acid sequences of proteins (Spencer and Nowick 2015; Neumann-Staubitz and Neumann 2016). The atoms of some amino acids generate distinctive X-ray scattering and magnetic resonance patterns. Structural biologists take many X-ray images or nuclear magnetic resonance (NMR) readings of proteins whose linear sequences differ by just a few amino acids (Mitchell and Gronenborn 2017). Comparing X-ray diffraction patterns produced by slightly different proteins helps scientists overcome experimental challenges like the phase problem in X-ray crystallography (Barwich 2017).[15]

So far, my discussion has shown that protein coding genes are useful for engineering novel biomolecules and producing distinctive observable effects, but are they explanatory? Biologists often do single out protein coding genes to explain the effects they produce in experimental situations (Baxter 2019). Differences in the nucleic acid sequences of fluorescent proteins and proteins used in X-ray crystallography or NMR spectroscopy can (at least sometimes) cause differences in an observable effect—like the color of fluorescence, intensity of X-ray scattering or NMR readings. Biologists often rely on observable effects to make inferences about otherwise unobservable phenomena. When biologists are called upon to justify the inferences they make about a target phenomenon, they appeal to differences in the nucleic acid sequences of protein coding genes to explain how the observable effect was produced. At least in experimental contexts, biologists care very much about explaining phenomena that may not have evolved by natural means or even exist outside the laboratory. In this way, the explanations biologists formulate that appeal to the causal significance of protein coding genes is narrow in scope in the populations and situations in which the explanation applies.

Yet protein coding genes are also of explanatory significance when it comes to some phenotypic traits as well. Consider Huntington's disease. Huntington's disease is a neurodegenerative monogenetic disorder caused by specific mutations in a single dominant protein coding gene. The huntingtin mutation consists of an expanded number of unstable CAG (cytosine-adenosine-guanine) repeats. Individuals carrying a single huntingtin gene with about forty or more CAG repeats will have nearly a 100 percent chance of developing the disease in all cases (Ross 2023; Arévalo, Wojcieszek, and Conneally 2001). Moreover, the number of CAG repeats is thought to inversely associate with the age of onset, with higher numbers of repeats causing earlier onsets. When it comes to the occurrence/nonoccurrence of Huntington's disease, it is the presence/absence of the huntingtin mutation that makes the difference. That is, the presence or absence of an expanded number of CAG repeats in the huntingtin gene. In fact, the huntingtin mutation may even be a fine-grained difference maker when it comes to the age of onset. Typically, more than seventy CAG repeats associates with juvenile onset, whereas between forty and seventy repeats associates with adult onset. What is noteworthy about this case is that the explanatory scope that the huntingtin gene has is broader than the explanatory scope of the protein coding genes previously mentioned. Huntington's disease is a highly penetrant disease—every individual carrying the huntingtin mutation develops the disease. This means that differences in the huntingtin gene account for differences in the occurrence/nonoccurrence of the disease for nearly all human populations. Importantly, explanations of Huntington's disease that appeal to huntingtin mutations are only partial. Such explanations hardly explain everything we might want to know about Huntington's disease—like how the disease originated and how it has been maintained in human populations. Moreover, diseases like Huntington's are very much an exception. Most genetic diseases have multiple—sometimes hundreds—of genetic determinants, each of which has a very low penetrance. Nevertheless, it remains that protein coding genes can (at least sometimes) explain phenotypic traits.

The discussion of this section demonstrates that the explanatory scope of protein coding genes is not uniform, as Waters and Weber have claimed. The metaphysical picture that falls out of their claim that protein coding genes only explain the linear sequences of RNA and proteins is one where the scope of each protein coding gene's explanatory and investigative significance is the same. I have argued in this section that things are not this neat. The explanatory and investigative significance of protein coding genes extends significantly beyond the linear sequences of RNA and proteins. They can be used for a very heterogeneous set of things—ranging from the manipulating the color of fluorescence, X-ray diffraction and NMR reading patterns, and even some phenotypic traits. Importantly, the scope of investigative and explanatory significance that one protein coding gene has will differ from another. For example, the huntingtin gene may have a broader explanatory and investigative significance than, say, a protein coding gene that encodes unnatural amino acids that is used for X-ray crystallography (Liu and Schultz 2010).

5. THE DAPPLED WORLD OF GENETICS—CONCLUSION

The world of genetics is a mess. It is even messier than philosophers have appreciated. Waters has argued that two main investigative and theoretical frameworks have characterized genetics in the past few centuries—the classical and molecular protein coding gene frameworks. Waters and Weber argue further that these frameworks offer partial, local explanations. In this chapter, I have argued that this view is neither sufficiently pluralist nor sufficiently general to fully capture the mess that constitutes the world that geneticists investigate and explain. It is not sufficiently pluralist because there are many molecular gene concepts at play in addition to the molecular coding gene concept. In addition to the molecular coding gene concept, there are molecular gene concepts that count regulatory gene sequences as a proper part of the gene, while others treat *cis*-regulatory sequences as distinct genes. In contrast to the molecular coding gene concept, these other gene concepts account for different types of phenomena and serve different investigative purposes. It is not sufficiently general, because protein coding genes can (sometimes) explain differences in phenomena that extend beyond the linear sequences of RNA and proteins. Protein coding genes are amenable to direct manipulation by researchers. Researchers continue to get better and better at precisely determining the nucleic acid sequences of protein coding genes. This has made protein coding genes especially useful "handles" by which to manipulate and control life processes for experimental and technological purposes. Beyond the artificial confines of the laboratory, differences in protein coding genes can also account for differences in some phenotypic traits.

Much of the metaphysical view that Waters and Weber advance is left unchallenged by this argument; however, what the arguments of this chapter do is complicate how we should characterize the messy nature of genetics. Explanations that appeal to regulatory or protein coding genes are partial and limited in scope. Genes aren't a fundamental entity of the world but especially useful tools for manipulating and controlling life processes. However, Waters and Weber are drawing (1) too few boundaries and (2) boundaries in the wrong places. They are drawing too few boundaries because they have overlooked the explanatory and investigative significance of alternative molecular gene concepts. There are in fact many more ways to "carve up" the world of contemporary genetics than just in terms of the molecular coding and classical gene concepts. They are also drawing their boundaries in the wrong places when it comes to the investigative and explanatory scope of protein coding genes. They suggest that the explanatory scope extends only to the linear sequences of RNA and protein coding genes. This is mistaken. The explanatory significance of protein coding genes often extends significantly beyond the linear sequences of RNA and protein coding genes. However, there is no simple way to characterize the explanatory scope of protein coding genes. Different protein coding genes will have different explanatory scope. This is because the explanatory scope of a gene (protein coding or otherwise) depends on a myriad of factors, such as the investigative interests of scientists at a point in history, the penetrance of a gene, genetic diversity in a population, the investigative techniques and tools that are available to scientists at a time, and more. Notice that this list of factors includes both objective features of the world and epistemic features of scientists. At least some of these factors (epistemic features of scientists) will change over history as new techniques and tools are developed. In turn, this can change the explanatory scope of a given protein coding gene. For example, the explanatory scope of the gene for green fluorescent protein was narrower than it is today before it was developed into an observational tool. This means that the boundaries that demarcate the explanatory and investigative significance will differ for each gene and will likely be irregularly shaped and sized.

The picture of reality that emerges from this discussion is very much in the image of Nancy Cartwright's (1999) dappled world notion. Cartwright's view is aimed specifically at physics and economics, but it is applicable to the area of genetics as well. In her view, the empirical success of our best theories of physics and economics suggest their truth but not their universality—far from it. The empirical successes of physics and economics are extremely limited in scope and confined to very specific experimental conditions—"arranged *just so*" (Cartwright 1999, 2). The dappled world notion describes patches of order and regularity. The patches are not uniform; they take irregular shapes and sizes. Much the same can be said about the world of genetics. In many cases, with enormous labor and thought scientists can arrange living systems in just the right configuration so that genes can exert causal regular control over a host of processes. This is true in experimental situations as well as in the management of genomic databases. It takes a lot of painstaking work to prepare proteins for X-ray crystallography, and it has taken even more labor to develop synthetic technologies that help crystallographers solve the phase problem. Thus, the explanations crystallographers formulate to justify the inferences they make about a protein's structure are restricted to the experimental conditions they engineer. It takes a different kind of labor to facilitate manipulation and control of life processes by managing genomic databases. NCBI's GenBank employs its distinctive molecular gene concept to organize and structure classificatory theories in such a way that facilitates inquiry. Just as the explanations in protein crystallography are restricted to specific experimental conditions, classificatory theories in a genomic database are often restricted to a limited range of conceptual conditions embedded in the repository. Occasionally, some gene is a significant difference maker with respect to some life process beyond the walls of a laboratory, but this is likely to be an exception to the rule (as with monogenetic diseases). Either way the point is that the explanatory and investigative scope of genes constitute patches of regularity and order. Where the boundaries of each patch lie is likely to differ for different genes (types and tokens). An adequate understanding of where the boundaries lie not only enriches our picture of reality, it also acts as a guide to effective interaction with the world.

NOTES

1. Of course, it is quite common for many—sometimes hundreds of—genes to be singled out as significant difference makers for explanatory purposes (of, say, a phenotypic outcome or disease). In this case, each gene only has a very small degree of causal control over the outcome of interest and, thus, lies beyond the scope of this chapter.

2. The conceptual, theoretical, and investigative strategies of geneticists in other parts of the world at the time were importantly different from classical genetics in the United States. (For examples, see Goldschmidt 1928 and Harwood 1993.)

3. Some molecular coding genes only determine the linear sequences of RNA molecules. Thus, I employ the term "molecular coding gene" to describe cases where a segment of DNA determines the linear sequence of either an RNA or protein molecule.

4. There are a few additional amino acids (i.e., selenocysteine and pyrrolysine) commonly found in living systems on earth. Both selenocysteine and pyrrolysine are encoded by nonstandard means.

5. An average gene of, say, 900 nucleic acid bases will have 4^{900} possible alternatives, since there are four different nucleic acid bases possible for each position in the sequence.

6. Other biochemical processes have this property as well, such as frameshifting during protein translation (Griffiths and Stotz 2013; Falk 2010).

7. Of course, molecular coding genes needn't be identified with only the collection of nucleic acid sequences that make a difference to the linear sequences of gene products. I take this up in the following section.

8. For good analyses of the relationship between classical and molecular genetics, see Vance 1996 and Baetu 2011.

9. A gene concept I don't discuss but that may be yet another way to individuate molecular genes is Lenny Moss's (2003) Gene-P and Gene-D distinction.

10. Piatigorsky calls this the open gene concept. Piatigorsky (2007, 52–3) recognizes that the open gene concept can easily give way to something more like the postgenomic gene concept.

11. This is what Griffiths and Stotz (2006) call a nominal gene.

12. GenBank no longer assigns a gene ID to new entries. This change was due partly to gene ID numbers being redundant information as accession numbers are also assigned to gene entries and partly to the accession number being more "human-readable" (Benson et al. 2016).

13. Of course, there are reasons to be skeptical about Waters's inference from the ontology of a scientific paradigm to metaphysical conclusions about the world beyond paradigms. See Bausman 2023 for a thoughtful exploration of this problem. Personally, I am disinclined to make the kind of inference Waters employs. Rather, I prefer restricting myself to something like perspectival realism—or at least realism relative to scientific paradigms. Nevertheless, the point of this chapter is to argue that Waters's logic should prompt us to conclude something more radical than he admits.

14. The logic of this claim is compatible with there being more than one variable with the relevant explanatory property. In such cases, it would be principled and justified to single out all causal variables possessing the relevant explanatory property. So, in a case where there are multiple variables with fine-grained causal control, it would be appropriate to single out all such variables in explanation.

15. To make inferences about the structure of a protein, structural biologists need to introduce heavy atoms into their specimens. This has commonly been achieved by soaking proteins in a bath of heavy atoms; however, this method is limited by a lack of precision. Synthetic tRNA molecules have been engineered to incorporate amino acids carrying heavy atoms at site specific locations by synthetic biologists (Liu and Schultz 2010).

REFERENCES

Arévalo, J., J. Wojcieszek, and P. M. Conneally. 2001. "Tracing Woody Guthrie and Huntington's Disease." *Seminars in Neurology* 21, no. 2: 209–23.

Baetu, T. M. 2011. "The Referential Convergence of Gene Concepts Based on Classical and Molecular Analyses." *International Studies in the Philosophy of Science* 24, no. 4: 411–27.

Baetu, T. M. 2012a. "Genes after the Human Genome Project." *Studies in History and Philosophy of Biological and Biomedical Sciences* 43: 191–201.

Baetu, T. M. 2012b. "Genomic Programs as Mechanism Schemas: A Non-Reductionist Interpretation." *British Journal for the Philosophy of Science* 63, no. 3: 649–71.

Barwich, S. A. 2017. "Is Captain Kirk a Natural Blonde? Do X-Ray Crystallographers Dream of Electron Clouds? Comparing Model-Based Inferences in Science with Fiction." In *Thinking about Science, Reflecting on Art,* edited by O. Bueno, G. Darby, S. French, and D. Rickles. New York: Routledge.

Bausman, W. C. 2023. "How to Infer Metaphysics from Scientific Practice as a Biologist Might." In *From Biological Practice to Scientific Metaphysics,* edited by W. C. Bausman, J. K. Baxter, and O. M. Lean. Minneapolis: University of Minnesota Press.

Baxter, J. 2019. "How Biological Technology Should Inform the Causal Selection Debate." *Philosophy, Theory, and Practice in Biology* 11, no. 002: 1–17.

Benson, D. A., I. Karsch-Mizrachi, D. J. Lipman, J. Ostell, and D. L. Wheeler. 2007. "GenBank." *Nucleic Acid Research* 35: D25–D30.

Benson, D. A., M. Cavanaugh, K. Clark, I. Karsch-Mizrachi, D. J. Lipman, J. Ostell, and E. W. Sayers. 2016. "GenBank." *Nucleic Acid Research* 45: D37–D42.

Bridges, C. B., and T. H. Morgan 1919. "The Second-Chromosome Group of Mutant Characters." In *Carnegie Institution of Washington Publication* 278: 123–304.

Burian, R. 2004. "Molecular Epigenesis, Molecular Pleiotropy, and Molecular Gene Definitions." *History and Philosophy of the Life Sciences* 26: 59–80.

Cartwright, N. 1999. *The Dappled World: A Study of the Boundaries of Science.* New York: Cambridge University Press.

Chalfie, M., Y. Tu, G. Euskirchen, W. Ward, and D. Prasher. 1994. "Green Fluorescent Protein as a Marker for Gene Expression." *Science* 263: 802–5.

Dupré, J. 1993. *The Disorder of Things: Metaphysical Foundations of the Disunity of Science.* Cambridge, Mass.: Harvard University Press.

Englesberg, E., and G. Wilcox. 1974. "Regulation: Positive Control." *Annual Review of Genetics* 8: 219–42.

Falk, R. 2010. "What Is a Gene? Revisited." *Studies in History and Philosophy of Biological and Biomedical Sciences* 41: 396–406.

Gerstein, M. B., C. Bruce, J. S. Rozowsky, D. Zheng, J. Du, J. O. Korbel, O. Emanu-elsson, Z. D. Zhang, S. Weissman, and M. Snyder. 2007. "What Is a Gene, Post-ENCODE? History and Updated Definition." *Genome Research* 17: 669–81.

Goldschmidt, R. 1928. "The Gene." *The Quarterly Review of Biology* 3, no. 3: 307–24.

Griffiths, P., and E. M. Neumann-Held. 1999. "The Many Faces of the Gene." *Bioscience* 49, no. 8: 656–62.

Griffiths, P., and K. Stotz. 2006. "Genes in the Postgenomic Era." *Theoretical Medicine and Bioethics* 27: 499–521.

Griffiths, P., and K. Stotz. 2013. *Genetics and Philosophy: An Introduction.* New York: Cambridge University Press.

Gutierrez-Preciado, A., M. Peimbert, and E. Merino. 2009. "Genome Sequence Database: Types of Data and Bioinformatic Tools." In *Encyclopedia of Microbiology,* edited by Moselio Schaechter, 211–36. Oxford: Elsevier.

Harwood, J. 1993. *Styles of Scientific Thought: The German Genetics Community 1900–1933.* Chicago: University of Chicago Press.

Havstad, J. C. 2016. "Proteins, Tokens, Types, and Taxa." In *Natural Kinds and Classification in Scientific Practice,* edited by C. Kendig. New York: Routledge.

Havstad, J. C. 2017. "Messy Chemical Kinds." *British Journal for the Philosophy of Science* 69, no. 3: 719–43.

Jacob, F. 1977. "Genetics of the Bacterial Cell." In *Nobel Lectures,* edited by D. Baltimore. New York: Elsevier North Holland.

Jacob, F. 1988. *The Statue Within: An Autobiography.* New York: Basic Books.

Jacob, F., and J. Monod. 1961a. "Genetic Regulatory Mechanisms in the Synthesis of Proteins." *Journal of Molecular Biology* 3: 318–56.

Jacob, F., and J. Monod. 1961b. "On the Regulation of Gene Activity." *Cold Spring Harbor Symposia on Quantitative Biology* 26: 193–211.

Jinek, M., K. Chylinski, I. Fonfara, M. Hauer, J. A. Doudna, and E. Charpentier. 2012. "A Programmable Dual-RNA-Guided DNA Endonuclease in Adaptive Bacterial Immunity." *Science* 337, no. 6096: 816–21.

Judson, H. F. 1996. *The Eight Day of Creation: Makers of the Revolution in Biology.* New York: Cold Spring Harbor Laboratory Press.

Kanehisa, M., J. W. Fickett, and W. B. Goad. 1984. "A Relational Database System for the Maintenance and Verification of the Los Alamos Sequence Library." *Nucleic Acids Research* 12, no. 1.

Keller, E. F. 2002. *The Century of the Gene.* Cambridge, Mass.: Harvard University Press.

Kohler, R. E. 1994. *Lords of the Fly: Drosophila Genetics and the Experimental Life*. Chicago: University of Chicago Press.

Leonelli, S. 2016. *Data-Centric Biology: A Philosophical Study*. Chicago: University of Chicago Press.

Lettice, A. L., S. J. H. Heaney, L. A. Purdie, L. Li, P. de Beer, B. A. Oostra, D. Goode, G. Elgar, R. E. Hill, and E. de Graaff. 2003. "A Long-Range Shh Enhancer Regulates Expression in the Developing Limb and Fin and Is Associated with Preaxial Polydactyly." *Human Molecular Genetics* 12, no. 14: 1725–35.

Liu, C. C., and P. G. Schultz. 2010. "Adding New Chemistries to the Genetic Code." *Annual Review of Biochemistry* 79: 413–44.

Machamer, P. K., L. Darden, and C. F. Craver. 2000. "Thinking about Mechanisms." *Philosophy of Science* 67: 1–25.

McConwell, A. K. 2017. "Contingency and Individuality: A Plurality of Evolutionary Individuality Types." *Philosophy of Science* 84, no. 5: 1104–16.

Mitchell, S. D., and A. M. Gronenborn. 2017. "After Fifty Years, Why Are Protein X-Ray Crystallographers Still in Business?" *British Journal for the Philosophy of Science* 68: 703–23.

Moss, L. 2003. *What Genes Can't Do*. Cambridge, Mass.: MIT Press.

Neumann-Staubitz, P., and H. Neumann. 2016. "The Use of Unnatural Amino Acids to Study and Engineer Protein Function." *Current Opinion in Structural Biology* 38.

Piatigorsky, J. 2007. *Gene Sharing and Evolution*. Cambridge, Mass.: Harvard University Press.

Portin, P. 2009. "The Elusive Concept of the Gene." *Hereditas* 146: 112–17.

Ross, L. N. 2023. "Explanation in Contexts of Causal Complexity: Lessons from Psychiatric Genetics." In *From Biological Practice to Scientific Metaphysics*, edited by W. C. Bausman, J. K. Baxter, and O. M. Lean. Minneapolis: University of Minnesota Press.

Sayers, E. W., J. Beck, J. R. Brister, E. E. Bolton, K. Canese, D.C. Comeau, K. Funk, et al. 2020. "Database Resources of the National Center for Biotechnology Information." *Nucleic Acids Research* 48: D9–D16.

Schaffner, K. F. 1993. *Discovery and Explanation in Biology and Medicine*. Chicago: University of Chicago Press.

Shen, Y., Y. Chen, J. Wu, N. C. Shaner, and R. E. Campbell. 2017. "Engineering of mCherry Variants with Long Stokes Shift, Red-Shifted Fluorescence, and Low Cytotoxicity." *PLOS One* 12, no. 2: 1–14.

Singer, M., and P. Berg. 1991. *Genes and Genomes: A Changing Perspective.* Mill Valley, Calif.: University Science Books.

Spencer, R. K., and J. S. Nowick. 2015. "A Newcomer's Guide to Peptide Crystallography." *Israel Journal of Chemistry* 55: 698–710.

Stotz, K. 2004. "With 'Genes' Like That, Who Needs an Environment? Postgenomic's Argument for the 'Ontogeny of Information.'" *Philosophy of Science* 73, no. 5: 905–17.

Stotz, K. 2006. "Molecular Epigenesis: Distributed Specificity as a Break in the Central Dogma." *History and Philosophy of the Life Sciences* 28: 527–44.

Vance, R. E. 1996. "Heroic Antireductionism and Genetics: A Tale of One Science." *Philosophy of Science* 63: S36–45.

Wain, H. M., E. A. Bruford, R. C. Lovering, M. J. Lush, M. W. Wright, and S. Povey. 2002. "Guidelines for Human Gene Nomenclature." *Genomics* 79, no. 4: 464–70.

Waters, C. K. 1994. "Genes Made Molecular." *Philosophy of Science* 61: 163–85.

Waters, C. K. 2004. "What Was Classical Genetics?" *Studies in History and Philosophy of Science* 35: 783–809.

Waters, C. K. 2006. "A Pluralist Interpretation of Gene-Centered Biology." In *Scientific Pluralism, Minnesota Studies in the Philosophy of Science, Volume XIX*, edited by S. H. Kellert, H. E. Longino, and C. K. Waters. Minneapolis: University of Minnesota Press.

Waters, C. K. 2007. "Causes That Make a Difference." *Journal of Philosophy* 104, no. 11: 551–79.

Waters, C. K. 2010. "Beyond Theoretical Reduction and Layer-Cake Antireduction: How DNA Retooled Genetics and Transformed Biological Practice." In *The Oxford Handbook of Biology*, edited by Michael Ruse. Oxford: Oxford University Press.

Waters, C. K. 2017. "No General Structure." In *Metaphysics and the Philosophy of Science: New Essays,* edited by M. H. Slater and Z. Yudell. New York: Oxford University Press.

Waters, C. K. 2019a. "Ask Not 'What *Is* an Individual?'" In *Individuation, Process, and Scientific Practices*, edited by O. Bueno, R.-L. Chen, and M. B. Fagan. Oxford: Oxford University Press.

Waters, C. K. 2019b. "Presidential Address, PSA 2016: An Epistemology of Scientific Practice." *Philosophy of Science* 86: 585–611.

Weber, M. 2006. "The Central Dogma as a Thesis of Causal Specificity." *History and Philosophy of Life Sciences* 28: 595–610.

Weber, M. 2013. "Causal Selection vs Causal Parity in Biology: Relevant Counterfactuals and Biologically Normal Interventions." In *Causation in Biology and Philosophy*, edited by C. K. Waters, M. Travisano, and J. Woodward. Minneapolis: University of Minnesota Press.

Weber, M. 2017. "Discussion Note: Which Kind of Causal Specificity Matters Biologically?" *Philosophy of Science* 84, no. 3: 574–85.

Weber, M. 2023. "The Reduction of Classical Experimental Embryology to Molecular Developmental Biology: A Tale of Three Sciences." In *From Biological Practice to Scientific Metaphysics*, edited by W. C. Bausman, J. K. Baxter, and O. M. Lean. Minneapolis: University of Minnesota Press.

"Webinar: A Submitter's Guide to GenBank, Part 1: Using BankIt for Small-Scale Nucleotide Sequence Submission." 2014. https://youtu.be/OZxxsRm0pP4.

Wittkopp, P. J., and G. Kalay. 2012. "*Cis*-Regulatory Elements: Molecular Mechanisms and Evolutionary Processes Underlying Divergence." *Nature Reviews Genetics* 13: 59–69.

Woodward, J. 2010. "Causation in Biology: Stability, Specificity, and the Choice of Levels of Explanation." *Biology and Philosophy* 25: 287–318.

Wu, P. F., S. Guo, X. F. Fan, L. L. Fan, J. Y. Jin, J. Y. Tang, and R. Xiang. 2016. "A Novel ZRS Mutation in a Chinese Patient with Preaxial Polydactyly and Triphalangeal Thumb." *Cytogenetic and Genome Research* 149, no. 3: 171–75.

7

THE REDUCTION OF CLASSICAL EXPERIMENTAL EMBRYOLOGY TO MOLECULAR DEVELOPMENTAL BIOLOGY
A Tale of Three Sciences

MARCEL WEBER

I ATTEMPT TO CHARACTERIZE the relationship of classical experimental embryology (CEE) and molecular developmental biology and compare it to the much-discussed case of classical genetics. These sciences are treated here as discovery practices rather than as definitive forms of knowledge. I first show that CEE had some causal knowledge and hence was able to answer specific why-questions. A paradigm was provided by the case of eye induction, perhaps CEE's greatest success. The case of the famous Spemann-Mangold organizer is more difficult. I argue that before the advent of molecular biology, knowledge of its causal role in development was very limited. As a result, there was no functional definition of the concept of organizer. I argue that, like the classical gene concept, it is best viewed as an operational concept. This means that an account of reduction such as Kim's functional reduction, which is still a mainstay in scientific metaphysics, cannot work in these cases. Nonetheless, again like in the classical gene case, the operational concepts of CEE played an important heuristic role in the discovery of molecules involved in morphogenesis and cell differentiation. This was made possible by what I call inter-level investigative practices. These are practices that combine experimental manipulations targeting two (or more) different levels. I conclude that the two sciences are more closely related via their experimental practices than by any inter-level explanatory relations.

1. INTRODUCTION

Debates about reduction, mechanism, and physicalism have accompanied developmental biology since its historical beginnings in the nineteenth century (Weber 2022). While some biologists and philosophers thought that development can only be explained by postulating immaterial vital forces, the recent identification of numerous molecular mechanisms involved in the control of cell differentiation and growth seems to support some form of reductionism or another (Rosenberg 2006). Indeed, it is a recurrent feature of modern biology that a once-autonomous discipline with its own concepts, theories, investigative techniques, and so on becomes absorbed into the molecular mainstream and starts to use molecular techniques to identify and characterize molecules that play a role in the phenomena they are interested in.[1] Scientists and philosophers alike have given in to the temptation of describing this trend toward molecularization in terms of reduction and seeking parallels to other sciences, in particular physics. According to the standard view, reduction in physics consists of a derivation of the laws of the theory to be reduced from a fundamental theory.[2] Attempts by philosophers of biology to apply this model to the case of classical genetics and molecular biology (Schaffner 1969) eventually led to a consensus according to which this is not possible because the necessary bridge principles connecting the terms of the two theories do not exist (Hull 1974; Rosenberg 1985). In a nutshell, this is because central genetic concepts such as dominance can be realized by many different molecular mechanisms, making it impossible to find a molecular concept that corresponds to it exactly, which is what reduction traditionally conceived would require.[3] Thus, physics turned out to be a poor model for reduction in biology.

Since then, philosophers of biology, as well as some physicalistically oriented philosophers of mind, have shifted attention from theory reduction to reductive explanation, asking not if and how some central theory might have been reduced to a fundamental theory but instead focusing on cases where some specific biological phenomenon such as the brain's ability to perform a cognitive task or muscle tissue's ability to contract has been explained by identifying a molecular mechanism, or more generally a mechanism at a lower level (Wimsatt 1976; Weber 2005; Kim 2007; Craver 2007; Kaiser 2015; Bechtel 2006).

While these accounts are compelling at least for some cases, they mainly concern what is known as *inter-level reduction* (i.e., the question of how the properties of complex systems can be explained by the properties of the parts). This raises the question if cases of *diachronic* reduction (Nickles 1973), such as the historical transition of a body of knowledge like classical genetics to molecular biology, can also be analyzed in terms of reductive explanation.

Before we can answer this question, we must first know what should be reduced to what. In the much-discussed case of classical genetics and molecular biology, it turned out to be difficult to even answer this first question. Inspired by physics, we might look first for an explanatory theory or set of principles of classical genetics and then for a fundamental theory that explains it. The most obvious candidate for such an explanatory theory is the account of various inheritance patterns that classical geneticists offered. Here is an example: the fact that two genes are located on different chromosomes explains why they assort independently in Mendelian crosses, while linkage in genetic crosses is explained by the genes being located close together and on the same chromosome. Occasional failure of linkage is explained by the chromosomes sometimes exchanging parts during their alignment in meiosis, which can be observed under a light microscope. Thus, the pairing and separation of chromosomes during meiosis explains the regularities of gene transmission discovered by classical geneticists. Some philosophers, most notably Kitcher (1984), have argued that this explanation, which remains strictly at the cytological level, is better than any molecular account, which would provide only "gory details."

Waters (2008) argues that this view is mistaken in several ways: First, Kitcher was unduly pessimistic about molecular biology's potential to discover the molecular mechanisms behind meiosis. Second, these molecular accounts do improve over the purely cytological accounts of these processes. However, seeking and finding these better explanations is not the development that has transformed genetics. In fact, it is just one among many achievements, and a more peripheral one for that matter. What transformed biology is what Waters calls a new basic theory that contains a new understanding of what genes are as well as explanations of how DNA is replicated and how proteins are made. However, this basic theory is unlike the fundamental theories of physics in that it is not able to explain all the facts in its domain. Instead, the basic theory is best viewed as being part of a new toolbox that still contains methods from classical genetics and that is being used by biologists to learn more and more about the functions of various parts of the organism.[4] This is what the molecular revolution in biology is all about. For example, a mutant of the nematode worm *C. elegans* named *unc*-70 (originally discovered by using methods from classical genetics) allowed biologists to unravel the function of the protein β-spectrin in the development of the nervous system.[5]

Waters's account shows that an exclusive focus on theories and explanations has led philosophers to misconstrue the relationship between classical genetics and molecular biology. Close attention to the practices of discovery used in these disciplines has led to a more adequate characterization of how these sciences are related.

In this chapter, I would like to extend this discussion to an area that has received much less attention from philosophers than classical genetics, namely classical experimental embryology (henceforward CEE) from the early twentieth century and its relationship to molecular developmental biology, a field that has seen major advances in recent decades. CEE was initially mainly practiced on amphibian embryos, which are easy to manipulate and have the advantage of developing in isolation. One of CEE's chief interests concerned the phenomenon of induction. This designates a process during which embryonic tissue that would by default follow some specific developmental path receives a signal from another tissue that changes its fate (e.g., from normal epidermis to lens tissue). It has been suggested that CEE did not provide any explanations of such phenomena; it merely described them, while genuine explanations had to await the discovery of the first molecules that bring about these changes, for example, so-called homeobox genes and their products (Rosenberg 1997). In this view, all that CEE contributed was merely a set of explananda and no explanantia nor anything else.

I will draw a somewhat different picture of how CEE and molecular developmental biology are related. By examining the classic examples of eye induction and Spemann's organizer, I will show that CEE did have some causal and hence explanatory knowledge. Perhaps the clearest example is provided by the case of eye induction. However, I will also show in section 2 that a particularly influential finding from CEE, the Spemann-Mangold experiment (including subsequent refinements thereof), provided at most a very weak kind of causal knowledge.

I will show in section 3 that this makes it difficult to assimilate this case into an account of explanatory reduction such as Kim's (2007). I will argue that, much like the classical concept of the gene, the organizer is better understood as an operational concept with enormous heuristic value rather than a functional concept with mainly explanatory value. Thus, CEE provided more than just explananda (phenomena to be explained) and perhaps some modest causal knowledge. It provided also investigative techniques and operational concepts that served as important tools for identifying molecules. In section 4, I will propose that CEE became part of what I call an *inter-level investigative practice* in which certain classical concepts functioned as research tools. In section 5, I will draw together the parallels and differences to the case of classical genetics.

Without further ado, I will now take a look at CEE by paying special attention to the question of what kind of knowledge this science had.

2. THE KNOWLEDGE OF CLASSICAL EXPERIMENTAL EMBRYOLOGY

The origins of CEE are often identified with the German tradition of *Entwicklungsmechanik* that emerged mid-nineteenth century. Wilhelm Roux is widely seen as the founder of this tradition, but as Maienschein (1991) shows, there was increasing interest in individual development and in experimental methods in various research centers at that time, not only in Germany. Nonetheless, Roux's experiments with frog embryos were clearly influential. Like many embryologists at that time, Roux was interested in the causes of cell differentiation in the first cleavage divisions of amphibian embryos. The initial debate was about the question of whether these causes are internal or external to the embryo. An example of an external factor would be gravity, but unlike in plants it seemed to have no effect on animal development. Roux's preferred theory was one of "self-differentiation" according to which inner causes determined the first cells toward different fates. One of his most influential experiments involved the destruction of one of the cells of a two-cell frog embryo by punctuation. He found that the surviving cell will typically form a partial embryo and took this to support his "mosaic theory" of development according to which the different parts of the embryo develop independently (Roux 1888).

Another influential investigation was carried out by Hans Driesch on sea urchin embryos. Driesch found that each one of the cells of a sea urchin up to the eight-cell stage was able to form a whole (if smaller) pluteus larva (Driesch 1892). He interpreted this phenomenon in terms of regulation.[6] In his view, embryonic cells were internally determined to the same fate, but they were able to change their differentiation state as a function of their surroundings. He introduced a distinction between the prospective fate (*prospektive Bedeutung*) and prospective potency (*prospektive Potenz*), where the former is a function of an embryonic part's location within the whole embryo. A region with constant prospective potency forms a harmonic-equipotential system (*harmonisch-äquipotenzielles System*). Driesch later used this phenomenon in his arguments for vitalism (Weber 1999), but at that time, he was still an orthodox *Entwicklungsmechanist*. What his findings showed was that the fate of an embryonic cell was not predetermined by some rigid program. Rather, the cells take cues from the surrounding cells when they divide and adapt their differentiation state according to their position in the embryo. A rather spectacular example is provided by the two blastomeres of the two-cell sea urchin embryo: when they are attached to each other, each of these cells will form a half embryo. When detached, each one can form a whole embryo.

One of the most intensely studied phenomena—and perhaps CEE's most important explanatory success—was the induction of the lens by the neural tissue that will later become the brain. In 1901, Hans Spemann published a study where he destroyed the eye rudiment on one side of *Triturus* (Northern European newt) embryos with a hot needle and observed that the lens failed to form on that side where he had intervened (Spemann 1901). He thus concluded that lens formation in the epidermis was caused by the optic vesicle underlying the epidermis, a process that he termed "induction." This conclusion was also supported by experiments done by Warren H. Lewis (1904) showing that optic vesicles transplanted to the flank of frog embryos caused the appearance of a complete lens. Induction became one of the guiding concepts in CEE, and embryologists interpreted most of their results in these terms (Saha 1991). In other words, it served as some sort of a paradigm, perhaps even in Kuhn's (1970) sense. The simple induction account was later refined considerably in Spemann's and in other laboratories, telling a more nuanced causal story. However, the lens induction results as well as their theoretical interpretation turned out to be one of the most lasting contributions of CEE, and I will shortly provide reasons why it may be one of its greatest explanatory achievements.

Another highly influential finding was the famous Spemann-Mangold experiment published in 1924.[7] Spemann's PhD student Hilde Mangold cut a small piece of tissue from the upper blastopore lip (the place where invagination begins in the process of gastrulation) from *Triturus cristatus* embryos and transplanted it to the ventral side of another embryo of the closely related species *Triturus taeniatus*. She found that the transplant induced a secondary embryo on the embryo's ventral side. Spemann and Mangold thus introduced the idea of an "organizer," an embryonic tissue capable of changing the fate of recipient epidermis cells and organizing them into forming an entire new body axis as well as several rudimentary organs such as the neural tube and the ear vesicles. *Triturus cristatus* and *taeniatus* were deliberately chosen because these two species differ in pigmentation. This allowed Spemann and Mangold to show that the secondary embryo was mostly built from tissue of the recipient and that the donor tissue only gave rise to the notochord. Thus, it wasn't just the transplant growing into a secondary embryo; the transplant clearly did something to the recipient tissue to change its fate into forming a new body axis (instead of just the ventral epidermis that it would have formed without the intervention).

The Spemann organizer soon became the holy grail of CEE, not least because it held promise to isolate the substances that mediated the organizing effect (an approach not favored by Spemann, and indeed it failed, but not for the reasons that motivated Spemann's doubts; see Hamburger 1988). However, trouble with Spemann's organizer concept soon arose when it was shown that there are many so-called "heterologous inducers," substances other than a newt blastopore lip that had a very similar effect when inserted into the ventral side of a newt embryo. Such inducers included boiled organizer tissue, various fractions from such tissues, fatty acids and sterols, and even nonphysiological substances such as sand particles or methylene blue. In the axolotl, even a saline solution worked as an "organizer." These findings called the whole organizer concept into question because they suggested that the organizing power is really in the receptor tissue and that it was merely triggered by the intervention. Spemann always emphasized that the receptor tissue needs to be *competent* to be induced. Nonetheless, the general perception was that heterologous inducers were a "funeral march" for the organizer theory or even that "Spemann's organizer set developmental biology back by 50 years" (De Robertis 2006). Indeed, the exact interpretation of the organizer findings was controversial until very recently.

These problems notwithstanding, it would be a mistake to see CEE as a failure. The science clearly had explanatory achievements. At least it was able to answer some of the questions that its practitioners raised. In order to analyze what kind of knowledge CEE produced, I would like to come back to the paradigmatic case of lens induction. Here are some of the questions raised by Hans Spemann in his 1936 book:

[W]ie kommt dieses auffallende zeitliche und räumliche Zusammenpassen der einzelnen Entwicklungsprozesse zustande? Woher kommt es, daß die Linse gerade an derjenigen Stelle der Epidermis zu wuchern beginnt, wo sie vom Augenbecher berührt wird, gerade zu dem Zeitpunkt, wo die Anlage der Retina sich einbuchtet? Üben beiderlei Vorgänge einen Einfluss aufeinander aus [. . .]? Oder verlaufen vielmehr beiderlei Vorgänge unabhängig voneinander, unter Selbstdifferenzierung der getrennten Anlagen, und beruht ihr genaues Zusammenpassen auf einer vorher erfolgten genauen Abstimmung der Teile aufeinander? (Spemann 1936, 26)

[**My translation**] How does this remarkable temporal and spatial fit of the different developmental processes arise? Why does the lens start to grow in the very spot where the optical cup touches the epidermis, at the exact time when the retinal *Anlage* invaginates? Do both processes exert an influence on each other [. . .]? Or do they proceed independently of each other, under self-differentiation of the separate *Anlagen*, while their fit is based on a preexisting exact harmonization of the parts to each other?

As always when two events coincide in space and time, this could be due to a causal interaction or due to a preestablished harmony of processes that are causally isolated from each other. Spemann wondered which one it was in the case of eye development. His experiments strongly suggested that it was a causal interaction. This was an explanatory achievement: a question about causality—preestablished harmony versus interaction—was answered. By and large, this answer is still considered to be essentially correct today.[8]

Thus, in contrast to what Rosenberg (1997) claimed, CEE clearly had some causal knowledge, and this knowledge answered important why-questions like the ones about eye induction just discussed. I also believe that Driesch's findings briefly reviewed earlier told embryologists something important about developmental causality, namely that the developmental fate of cells can depend on causal interactions with other parts of the embryo. However, it must be admitted that all this causal knowledge was very rough in the sense that not many causal variables had been identified and the ones that were identified were not highly specific in the sense that they did not allow scientists to control developmental processes in a very fine-grained way.[9] Furthermore, what characterizes most of this knowledge is a certain remoteness of the causes from the effects.[10] This is particularly striking in the case of the organizer. The Spemann-Mangold experiment and its subsequent refinements merely showed that the blastopore lip (or parts of it) has the power of triggering the formation of a body axis in transplantation experiments. But what was actually observed was merely the end result of this reprogramming of the cells in the recipient tissue. Exactly what developmental events were triggered was unknown at the time. As I will show in the following section, the initial hypothesis (neural induction) turned out to be false.[11] Finally, the implications for normal development were not clear at the time, as witnessed by the extensive controversies surrounding the iconic organizer experiment.

In its remoteness of causes and effects, the case of the organizer resembles the classical gene, which was only known to cause uniform phenotypic differences in particular genetic and environmental contexts (called the "difference principle" of classical genetics; see Waters 2008 and section 3 later in this chapter).

If this is correct, why does the Spemann-Mangold experiment still feature so prominently in developmental biology textbooks? In order to answer this question, I suggest that we should look to the case of classical genetics and the way in which biological concepts sometimes serve as investigative tools.

3. OPERATIONAL CONCEPTS AND THE FAILURE OF THE FUNCTIONAL APPROACH TO REDUCTION

Since the 1980s, molecular developmental biologists have identified hundreds of genes and gene products in various organisms that are today believed to be responsible for some of the effects observed by early experimental embryologists. Many of these gene products are transcription factors, which means that they bind selectively to DNA at specific regions and activate or block the expression of specific genes. Others are signaling molecules that regulate the proliferation or differentiation of cells. Some of these molecules are secreted by the cells making up the organizer, diffuse through the embryonic tissue, and bind to receptors on other cells, which then send a signal to the nucleus, which changes the differentiation state of these cells. There are also molecules that regulate or cause cellular movements such as the invagination of the lens placode (a thickening of the epidermis that will later form the lens). Needless to say, these processes are enormously complex. Before considering to what extent knowledge about such molecules and their interactions reduces the knowledge about induction and organizers from CEE, let us first recall what such a reduction might look like.

It is clear that a model of reduction such as Ernest Nagel's (1961) according to which reduction consists in the derivation of some laws from more fundamental laws isn't applicable here. Even if we grant that the phenomena described by CEE can be stated in the form of laws, there is no molecular theory from which these laws can be deduced. Biological knowledge is usually not organized into theories that consist of a few basic principles that can in principle explain all the phenomena in its domain (Waters 2008).

A more promising approach to reduction is the one elaborated by Jaegwon Kim (2007) for his metaphysics of the mind, taking genetics and molecular biology as a model. According to Kim, a successful reduction proceeds as follows:

(1) Provide a functional definition of the phenomenon to be reduced, i.e., having M =$_{def}$ having some property or other P (in the reduction base domain) that exerts causal role C

(2) Identify the properties or mechanisms in the reduction base that perform C

(3) Explain how the realizers of M perform C

I will not consider the merits of this account in general; maybe there are positive examples to be found.[12] However, my contention is that such examples will concern at best cases of inter-level reductive explanation. Such cases differ considerably from putative cases of diachronic reduction, which concern the relationship between an older and a more recent body of knowledge (Nickles 1973). I suggest that Kim's account does not apply to such cases because some of the crucial scientific concepts do not admit of Kim-style functional definitions. If we consider the case of classical genetics, we will find that there was nothing like a causal role associated with the pre-molecular gene concept. Kim himself (2007, 101) suggests that the functional definition of a gene is "a mechanism that encodes and transmits genetic information." This is too broad, as there are many mechanisms that could be said to encode and transmit genetic information (Oyama 2000).[13] What is more, the notion of genetic information is notoriously unclear.

Thus, there is no such concept of the gene as Kim imagined it. There are only two things to be found in the practice of classical genetics: (A) There is what Waters (2008) calls the *difference principle*: differences in a classical gene cause uniform phenotypic differences in particular genetic and environmental contexts. As I have already argued in the last section, this is at best a remote or unstable causal link (which was nonetheless instrumentally important for identifying genes). (B) There are *operational criteria* that were used by classical genetics in order to identify genes experimentally. Briefly, these criteria involved very elaborate crossing experiments with a variety of different mutants of an organism. To put it in a nutshell, two (recessive) mutations were considered to affect different genes if the phenotypic effects disappeared when both mutations were present in the genome, a phenomenon known as "complementation." When the phenotypic effects did not disappear, they were considered to reside in the same gene. Of course, there are lots of complications with this so-called complementation test, but these need not concern us here. What matters is that classical genetics used a combination of genetic mapping and complementation test in order to assign mutations to genes and thus were able to localize genes and their approximate boundaries on the chromosomes.

I wish to claim that neither (A) nor (B) provides a functional definition of the gene in the sense of Kim. This is easy to see in the case of (A): To be a cause of uniform phenotypic differences in particular environmental and genetic contexts is not a unique property of genes. Furthermore, there is no one-to-one relation between phenotypes and genes.

What about (B)? I claim that this is no causal role of the kind that could feature in a functional definition either. There are two reasons. First, one would expect a causal role to be something that the entity in question exerts even if no experiment is performed. But complementation is an effect that shows up only in very specific experiments. Of course, the explanation for why complementation works has something to do with the genes' physiological causal role. But that role was unknown to classical geneticists, so it couldn't have been part of an explicit functional definition. The second reason is that there are many genes for which the complementation test doesn't work, and classical geneticists knew this. There are also cases where it gives completely misleading results because there are phenomena such as interallelic complementation, different mutations that cancel or partially cancel each other's effects even when they are located on the same gene. (Note that partial complementation is always an indication that something is odd.) All these phenomena were known to classical geneticists, and they didn't consider them as severe anomalies for the gene concept.

Another way of putting my claim is by saying that classical geneticists did not have a functional definition nor any other theoretical definition of the classical gene because all they had was an operational concept. By using the notion of operational concept, I am not committing to a form of operationalism according to which the meaning of all concepts is reducible to or even synonymous with a set of operations, as Bridgman (1927) famously suggested. I am not subscribing to a general philosophical view about meaning;[14] I only wish to commit to the following claims: First, classical genetics did not have a specification of a causal role that was necessary and sufficient for being a gene. Second, there were experimental procedures for detecting and localizing genes on chromosomes (procedures that broke down for some cases; see Weber 2005, ch. 7 for details). Thus, while the classical gene concept contained the criteria for experimentally identifying or detecting genes, it doesn't say what their causal role is in the organism's development. This doesn't mean that there was *no* theoretical knowledge about genes. For example, it was known that they are arranged linearly on the chromosomes. And of course it was known that differences in genes can cause differences at the phenotypic level (the "difference principle" according to Waters). But none of this theoretical knowledge amounts to a theoretical definition, since it was in itself insufficient for identifying genes. Thus, when I qualify the classical gene concept as "operational," what I mean is not that the concept had no theoretical content. I only mean that this content wasn't sufficient to identify genes.[15] We could say that classical genes are those entities that are arranged linearly on chromosomes, cause phenotypic differences, and are detectable by classical genetics' experimental techniques. But this is no theoretical definition; it contains an ineliminable reference to experimental techniques. Note also that there is no completed or corrected version of classical genetics that contains a functional definition. There may be a functional definition of the *molecular* gene, but that is supposed to occur only in step 2 of Kim's reduction model.

What is also important to note is that, even though there is no clean one-to-one mapping of classical and molecular genes, the operational criteria of classical genetics were extremely helpful for identifying lots of molecular genes (Weber 2004). Thus, what mattered most for the advancement of biology was not the existence of some explanatory relation between the theories or concepts of the classical and the molecular bodies of knowledge but the possibility of what I wish to call inter-level investigative practices, that is, practices that combine classical and molecular techniques. It is there that we find the most important diachronic relations, not by looking only at theories or at explanations. I will say more about this notion in section 4. But first, I wish to suggest that the case of inducers and organizers bears an important resemblance to the case of the classical gene.

Classical experimental embryologists led passionate debates as to what inducers or organizers are. They never reached a consensus (Hamburger

1988). More recently, some philosophers of biology have attempted to define the role of the Spemann organizer in abstract causal terms, arguing that it was initially conceived as a causally specific instructive cause but turned out to be merely a switch or a permissive cause (Calcott 2017; Bourrat 2019). While these analyses are conceptually illuminating, they cannot help explain why the organizer concept was and is so important in developmental biology. I would rather like to suggest that, much like the classical genes, organizers were basically *operationally* defined concepts. Evidence for this can be found in a recent review article by two contemporary biologists:

> The notion of an organizer refers to specific experiments that test the signalling ability of specific groups of cells in particular contexts. The use of the term "organizer" should therefore be restricted to the outcome of precise experiments: a heterologous allotransplant in the same embryo, similar to that performed by Spemann and Mangold. (Arias and Steventon 2018)

While this looks like a conception of the organizer that was corrected in hindsight with the benefit of recently acquired knowledge, I wish to claim that what Arias and Steventon refer to here is *one* sense of the term "organizer" that was always there in experimental embryology. Of course, although various biologists (Spemann and Waddington in particular) also used the term in a richer theoretical sense, a sense that imbues the dorsal lip tissue with some causal power or another, the only *uncontroversial* sense was an operational one. According to this sense, an organizer is just a tissue that has the causal power of changing the fate of recipient tissue to which it is grafted such that a new body axis formed there.[16] Thus, an organizer is defined operationally by this experimental test.

Furthermore, it had always been clear to Spemann that the host tissue must be competent to be induced; the formative power does not reside in the inducing tissue alone. However, his idea that the organizer caused neuralization of tissue that was initially committed to become epidermis turned out to be false. Spemann and other classical embryologists had always thought that the dorsal cells where the neural tube will form are initially committed to become epidermis and are induced to the neural pathway by the organizer. However, it was shown later that it was the other way around: the dorsal cells are committed for the nervous system from the beginning; what the organizer does is to antagonize signals that would change their fate to become epidermis (Gilbert and Barresi 2016, 355).

The upshot of this discussion is that there is no causal role that was un-controversially attributed to the organizer. By a "causal role," I mean some-thing like a specification of the concrete developmental events that the organizer causes, something like neural induction. As I have shown in the previous section, all that was really known was the end result of the orga-nizer's activity in transplantation experiments (i.e., a secondary embryonic axis formed). Insofar as it was accepted by the CEE community, the concept of organizer was operational; organizers in various species were detectable by certain transplantation experiments, and this is what defined them. How-ever, as I will show in the next section, these operational concepts were im-portant for scientific practice because they gave the scientists important hints where to look for the causes of morphogenetic processes. The case is similar to the case of the classical gene, which was also a purely operational concept and which eventually led biologists to the molecular genes. Thus, while an account of reduction such as Kim's works for neither case, the heuristic func-tion of these operational concepts is undeniable.

In the following section, I will show that there is nonetheless an impor-tant kind of inter-level relation between CEE and molecular biology, a rela-tion that is also present in classical and molecular genetics. But this is not a relation at the level of theoretical knowledge; it inheres rather in the prac-tice of CEE, classical genetics, and molecular biology.

4. THE ORGANIZER AND INTER-LEVEL INVESTIGATIVE PRACTICES

There is no doubt that since the heyday of CEE, molecular developmental biologists have made considerable progress in identifying some of the mol-ecules that are responsible for the phenomena discovered by the older experi-mental approach. Eye induction was shown to be mediated by gene products from the *Otx2, Pax6,* and *Lens1* genes, which seem to give epidermis cells the competence to induce lenses. When this tissue comes into contact with the optic vesicle, the genes *mafs, Soxs,* and *Prox1* are activated, which in turn activate the expression of crystallin genes needed to build the lens and the retina (Ogino and Yasuda 2000). Remarkably, strongly homologous genes are involved in eye development in the entire animal kingdom (Halder, Callaerts, and Gehring 1995).

In the case of the Spemann organizer, there is a whole plethora of secreted signaling molecules and transcription factors that were described since the mid-1980s. One of the more surprising findings was that many of the proteins produced by the organizer are growth-factor antagonists that compete with growth factors such as bone morphogenetic proteins (BMPs) for binding to their specific receptors. In the dorsal ectoderm (i.e., the region near the organizer), the effect of these BMPs is to block the neural developmental pathway to which these cells are committed. When the organizer becomes active, it secretes growth factor antagonists such as Noggin, Chordin, and Follistatin. These antagonists compete with the BMPs for their receptors and thus lift the block, thereby inducing the cells to become neural. It should be noted that before the age of the molecules, embryologists had thought that the dorsal cells are committed to become epidermis and what the organizer does is to change their fate to the neural pathway. According to molecular developmental biologists, this story is not correct: what the organizer does is to antagonize signals that commit the cells to become epidermis, thus allowing the neural default pathway to become active. Thus, the developmental role of the organizer tissue was falsely described. It is therefore not a case of functional reduction sensu Kim, according to which scientists discover the molecular realizers of previously known functional roles.

Just as an aside, it is my contention that the main reason why the molecular accounts are superior to the explanations provided by CEE is not that the molecular level is more "fundamental"—an obscure notion. Rather, it is the increased manipulability of the embryos and the developmental processes that comes with the molecular techniques. Molecular biology succeeds because it allows more different kinds of interventions (e.g., on genes, mRNA, proteins) as well as interventions that are closer to the ideal of a "surgical" intervention in the sense of Woodward (2003). An example would be the injection of a single mRNA species into an embryo (see later) as opposed to the grafting of a whole tissue. In addition, the causal links discovered were more direct. Some of them may also be more specific, which roughly means that the cause variables allow more fine-grained control over their effect variables (Weber 2006; 2017; forthcoming; Waters 2007; Woodward 2010; Griffiths et al. 2015). But these claims are not my focus in this chapter, and I shall not defend them here.[17] What I would like to consider instead is the role that the legacy of CEE played in making the molecular tools available for experimentation in the first place.

Before the 1980s, attempts to isolate some of the substances that might be responsible for processes like eye induction or the establishment of embryonic axes by the organizer failed, probably due to the extremely low concentrations in which these are present in embryonic cells. This situation changed dramatically with the advent of molecular techniques such as gene cloning and sequencing, transgenic organisms, or antisense RNAs (small RNAs that can specifically neutralize messenger-RNAs in the cell by forming double strands with them, thus rendering them inactive). With respect to developmental biology, two advances deserve special attention here:

(1) The cloning of the first developmental genes in the fruit fly *Drosophila* by brute-force approaches such as "walking on the chromosome," using the vast collections of available *Drosophila* mutants that show developmental abnormalities (Weber 2004). The first *Drosophila* genes cloned turned out to be extremely helpful for cloning developmental genes in other organisms as well (*Xenopus*, humans, mouse, zebrafish) because it turned out that they share highly conserved functional DNA elements such as the homeobox (Gehring 1998).

(2) The invention of cDNA cloning (the "c" stands for "complementary"), a technique that uses the enzyme reverse transcriptase isolated from retroviruses in order to make DNA copies of mRNAs (Maniatis et al. 1976). The DNA copies can then be inserted in bacterial plasmids for amplification, sequencing, and making transgenic organisms.

These two methods were instrumental for isolating some of the very first vertebrate genes implicated in the organizer phenomenon. For example, in 1992, the laboratory of Edward M. De Robertis cloned a gene called "*goosecoid*" (Cho et al. 1991). (The name is a fusion of the two *Drosophila* genes *gooseberry* and *bicoid*, which both show sequence homologies.) The technique they used is quite remarkable: They isolated messenger-RNA (mRNA) from the dorsal lip of the blastopore of *Xenopus*[18] gastrulae. From these mRNAs, they synthesized cDNA. Searching these cDNAs for sequence homologies to *Drosophila* homeobox-containing genes led them to a gene that is expressed specifically in the organizer region. Microinjection of *goosecoid* mRNA to the ventral region of *Xenopus* embryos mimicked the action of the Spemann-Mangold organizer (Blumberg et al. 1991; Cho et al. 1991; Robertis et al. 1992).

Another example is the cloning of the gene *noggin*. The laboratory of Richard M. Harland used a technique called expression cloning, making use of a repertoire of effects that were already known to classical embryologists: When treated with lithium chloride (LiCl) before gastrulation, amphibian embryos become "dorsalized," that is, all their cells form neural tissue. When the embryos are irradiated with UV, they become "ventralized," which is the opposite effect. Smith and Harland (1992) extracted mRNA from LiCl-treated cells and used it to construct a cDNA expression library. This means that the cDNA fragments were inserted into a bacterial plasmid that contained the necessary signals for any gene contained in it to be expressed. These plasmids were then injected into ventralized embryos to check for their ability to rescue the formation of dorsal mesoderm. Thus, a gene called *noggin* was isolated and shown to be expressed specifically in the organizer region in normal embryos, all over dorsalized embryos and not at all in ventralized embryos. This strongly suggested that *noggin* somehow helps controlling the pathway leading to neural development.

I will not be concerned with the wealth of molecular detail that was discovered subsequently, nor do I want to understand here what exactly these molecules and their interactions explain and how they explain it. As Waters has shown for the case of classical genetics (see section 1), we will miss all the action when we focus on explanatory relations alone. In my 2004 work, I have shown that the first genes involved in *Drosophila* development were isolated with the help of what I then called "hybrid techniques," techniques that combine methods from classical genetics and cytology with the new recombinant DNA technology. Here, I would like to introduce the idea of an inter-level investigative practice. By this, I mean practices that integrate experimental manipulations targeting different levels, such as the tissue or cellular and the molecular level.[19] Such practices played an important role in the early days of molecular developmental biology, for example, when the first *Drosophila* genes were cloned. For example, genes like *Antennapedia* or *Fushi tarazu* were first mapped genetically, using classical recombination mapping. Then DNA isolated from chromosome preparations was cloned and hybridized to giant chromosomes, using radioactively labeled DNA in order to visualize the chromosomal location on microscopic images of the giant chromosomes. This is an inter-level practice because higher-level structures, namely cytological chromosome preparations, were manipulated in the same experiment as micro-level entities, namely DNA.

The techniques used to clone some of the first *Xenopus* genes involved in the organizer phenomenon also constitute an inter-level practice. As we have seen, the blastopore lip tissue that revealed some of its causal powers in the classic Spemann-Mangold experiment was used to extract mRNA for cDNA cloning. This required the same kinds of manipulations as those applied by the classical embryologists. But it also required manipulations on the DNA molecules, for example, by treating the extracted mRNA with reverse transcriptase in order to synthesize cDNA. This combining of manipulations at different levels is what defines inter-level investigative practices.[20]

What is most important is that the success of this practice does not depend on there being reductive relations such as they were imagined by philosophers such as Kim's functional reduction using the concept of realization, Nagel's derivational reduction, or Putnam's (1975) and Kripke's (1980) a posteriori identity such as "water = H_2O." It also doesn't require that there be some kind of general structure of reality (Waters 2017). All that it takes is some significant overlap (no coextension or set inclusion) between the things picked out by the operational concepts (organizer, induction, gene) and the molecular components that are responsible for the phenomena in question. This appears to have been the case in genetics: some but not all of the map regions where classical techniques indicated the presence of a single gene were shown to contain a molecular gene (and, of course, many molecular genes could never have been found by classical techniques because they don't produce usable mutations). It is also true in the case of the organizer: the region identified by Spemann and Mangold roughly (but not exactly) corresponds to developmentally relevant regions of tissue-specific gene expression that have an effect on the fate of surrounding cells. The operational concepts used by the classical experimental sciences were sharp enough for playing an important heuristic role in research, but this doesn't support a more traditional kind of reduction or the metaphysical assumption of a general structure (Waters 2017).

5. A TALE OF THREE SCIENCES

As we have seen, there are considerable parallels between the case of classical genetics and classical experimental embryology, both in their relationship with molecular biology.

First, in both cases, there was a body of knowledge generated by experimentally manipulating living organisms. Both sciences discovered organismic parts that have certain causal powers, namely the power to cause phenotypic differences in defined environmental and genetic contexts (in the case of genes) and the power to change the fate of surrounding cells (in the case of the Spemann-Mangold organizer or the optic vesicle). However, not many causal variables were known at the time, and the causal relations between them were quite remote. In other words, what experimenters in the classical disciplines saw were merely the remote effects of highly complex webs of causal influence. Molecular biology identified some of the more proximate elements in these causal webs, such as transcription factors that bind to DNA and regulate gene activity. Thus, there were some weak explanatory relations between the older science and molecular biology.

Second, in both cases, there were entities for which there was a precise operational definition (i.e., protocols and criteria for experimentally identifying them) but no clear functional definition (i.e., a description of the entity's proximate causal role in the organism). Instead, there was a lot of theoretical speculation that wasn't part of a scientific consensus. If there was a scientific consensus, it covered at most ways of practicing the science and judgments about significant research problems.[21] Third, as a consequence of the previous point, there were no molecular realizers of previously known functional roles identified. Thus, there is no Kim-style functional reduction.

Fourth, there existed (and still exist) inter-level investigative practices that combine experimental manipulations on whole organisms or organism parts (e.g., embryonic tissues) with interventions at the molecular level. In the case of developmental biology, these inter-level practices were instrumental for identifying hundreds of proteins and protein-coding genes. Fifth, the molecular techniques widened the scientists' repertoire for targeted interventions that allow them to discover much more detailed networks of causal dependencies than was possible in the classical era of experimental embryology. The interventions enabled by molecular techniques were more surgical, there were more of them, and the causal dependencies discovered were less remote and more stable.

So there are considerable similarities between classical genetics and classical experimental embryology. However, there is also an obvious difference: genetics was and is a much more versatile tool for biological research than any of the techniques that were used in the classical era of embryology. Classical genetic methods, in particular the analysis of spontaneous mutants in a variety of different species, have been used with success not only in developmental biology but also in biochemistry, cell biology (e.g., Nurse 1975), behavioral biology (e.g., Konopka and Benzer 1971), and even evolutionary biology (e.g., Dobzhansky 1937). There is hardly any part of biology that has not been changed considerably by the techniques of classical genetics and the inter-level investigative practices to which it gave rise.

Another potential difference concerns the lower-level science (i.e., what would have been called the "reducing theory" in older discussions of reduction in science). As Ken Waters has shown, molecular biology provided a basic theory about how DNA as the genetic material is replicated and expressed. However, unlike so-called "fundamental" theories (as they are thought by some to exist in physics), this basic theory is not able nor does it aspire to explain all the phenomena in its domain. It only explains how DNA molecules can be copied to produce new DNA as well as RNA molecules with the same or a complementary sequence (including repair mechanisms), how RNA molecules are processed after transcription, how proteins are synthesized, and how these processes are regulated. This is a crucially important insight for understanding life processes, but it doesn't account for everything there is in biology (in the sense in which a unified field, if it existed, would account for all physical phenomena; see Weinberg [1992]). In any case, according to Waters, molecular biology's importance is not exhausted by the explanatory achievements of its basic theory. What is at least as important is the way in which it has expanded the biologist's repertoire for learning more about processes that lie outside the scope of the basic theory.

It is not a trivial task to find out if there is anything in molecular developmental biology that would correspond to Waters's basic theory of molecular biology, as the question if and in what sense developmental biology (or any other part of biology) might have theories has been controversially discussed.[22] It is not so clear what sort of thing deserves that honorific title and what is better just described as a hypothesis or a model. For the purposes of this chapter, nothing hinges on this question. What is clear is that molecular developmental biologists heavily use not only the experimental techniques from molecular biology but also the latter's knowledge about how genes are expressed and regulated, including processes such as RNA splicing, post-translational modifications, or DNA methylation, and how proteins can transmit signals within or between cells. This is the same as Waters's basic theory. In addition, they use knowledge that is more specific to animal development, for example, about proteins that mediate mechanical adhesion between cells. Thus, they use something like an extended basic theory.

What is important for the purposes of this chapter is that developmental biologists use the basic theory and its extensions not only for *explaining* developmental processes but also for *doing* things in the lab, in particular for designing experiments to learn more about these processes. A beautiful example of this is provided by the example of the expression cloning of the gene *noggin* that I briefly explained in section 4. In this technique, mRNA was first isolated from frog embryos. Then the enzyme reverse transcriptase was used to make DNA copies of these mRNAs. Finally, an in vitro protein-synthesis system was used to make the corresponding protein and check its biological activity in the developmental process. Theoretical knowledge and experimental techniques go hand in hand here to reveal causal effects of genes and proteins on the developmental process that were previously unknown. Thus, developmental biology was adapted to the universal toolbox of molecular biologists. Some of the older techniques such as Spemann's transplantation methods continued to be used at least for some time, much like in the case of classical genetics, but unlike the latter, they remained quite specific to the developmental biology of vertebrates.

ACKNOWLEDGMENTS

Versions of this chapter were presented at the Templeton Summer Institute "From Biological Practice to Scientific Metaphysics" in July 2018 (Taipei), the Fifth European Advanced School in the Philosophy of the Life Science in September 2018 at the KLI (Klosterneuburg), the Third International Conference of GWP, the German Society for Philosophy of Science in February 2019 (Cologne), and the Department of Philosophy, University of Salzburg, in March 2019. I would like to acknowledge helpful comments in particular from Elena Rondeau, Naïd Mubalegh, Ken Waters, Paul Hoyningen-Huene, Alan Love, Bill Wimsatt, William Bausman, Janella Baxter, Bengt Autzen, Tiberius Popa, Michael T. Stuart, Lorenzo Casini, Michal Hladky, Gregorio Demarchi, and Guillaume Schlaepfer.

NOTES

1. When I was an undergraduate student in biology in the 1980s, we were even told by one of our professors that whenever we find ourselves being bored in class, the most probable cause will be that the material presented to us makes no reference to molecules!

2. See Nagel 1961 for the canonical formulation and Dizadji-Bahmani, Frigg, and Hartmann 2010 for a recent defense of this view.

3. Classic examples of bridge principles include the equation relating temperature and mean kinetic energy in an ideal gas and entropy and the probabilities of finding a system in a set of defined microstates.

4. To my knowledge, this point was first made by Vance (1996).

5. In an earlier work, I gave a similar account of how *Drosophila* geneticists used the techniques and resources that came with their model organism

in order to identify the first molecular genes implicated in development (Weber 2004).

6. The difference between Roux's frog and Driesch's sea urchin results later turned out to be an experimental artifact created by Roux's method of punctuation. If the dead blastomere is properly removed from the embryo, which Roux didn't do, the frog and sea urchin embryos respond rather similarly to this kind of intervention (Maienschein 1991, 50).

7. An insider's account of Spemann's and Mangold's work by another student of Spemann's can be found in Hamburger 1988.

8. From its inception, Spemann's model of lens induction has been embroiled in controversy (Saha 1991). Some results indicated that the optic vesicle was not necessary to induce a lens. (So-called "free lenses" were observed repeatedly by several experimenters.) Furthermore, the experiments purporting to demonstrate the sufficiency of neural tissue to induce a lens in ectoderm were not entirely conclusive because it could not be ruled out that it was contaminated by ectoderm that was already committed. Indeed, this turned out to be the case. In the 1980s and onward, new methods for marking and tracing host and donor tissues in transplantation experiments (e.g., by using dyes or specific antibodies) allowed to determine more precisely at what stage the head ectoderm becomes competent for induction (Saha, Spann, and Grainger 1989; Grainger 1992). This work led to a multistep model of induction according to which both the neural tissue and the ectoderm receive several induction signals. Spemann's own view as articulated in his *Experimentelle Beiträge* monograph of 1936 was quite close to this model; however, Spemann may not yet have had the evidence for his view. Molecular studies done since the 1990s revealed a complex cascade of mutual interactions between the neural and ectodermal cell lineages that are now referred to as "mutual inductions" (Gilbert and Barresi 2016, 523; Ogino et al. 2012). This is considerably different from the initial idea of an asymmetric induction, but Spemann & Co. were right that there was causal interaction between the neural and ectodermal cell lineages.

9. This concept of causal specificity is due to Waters 2007 and Woodward 2010. In my 2022 work, I argue that Spemann defended the organizer concept by presenting it as a (somewhat) specific cause.

10. I hesitate to characterize this remoteness in terms of stability as Woodward (2010) does because I am not convinced that longer causal chains are necessarily less stable.

11. Indeed, De Robertis's remark about the Spemann-Mangold organizer "setting developmental biology back for 50 years" suggests that the famous experiment did not show much more than was already known to Driesch, namely that a group of cells' fate can be modified by the surrounding cells. It just showed this in a very dramatic way.

12. Hoyningen-Huene (1997) argues that pheromones provide a good example of a functional role for which a set of molecular realizers was identified. He emphasizes in particular that the realizers need not themselves constitute natural kinds, as many accounts of reduction require.

13. While Kim's account is designed to be able to deal with multiple realizations, this isn't an instance. To see this, compare it to the case of pain, which is thought to be multiply realizable. Thus, pain may be defined as having the property P_1 or P_2 or . . . or P_n in the reduction base domain such that the P_i exerts causal role C. The causal role of pain may be roughly described as the property of being caused by tissue damage and causing withdrawal behavior and screaming. However, note that, by definition, *all and only* the physical states that exert role C are pain. By contrast, not all states that can be said to encode or transmit genetic information are genes (e.g., DNA methylation states).

14. Sophisticated versions of operationalism about scientific concepts to which I am sympathetic have been worked out by Feest (2005; 2010) and by Chang (2007).

15. I am indebted to Bengt Autzen and Janella Baxter for pushing me to clarify this point.

16. In my 2022 work, I argue that organizer tissue supported causally specific interventions—in other words, interventions that allowed experimenters to control the outcome in a fine-grained way. This suggested *some* role in structuring the developmental process, but it wasn't clear what role exactly.

17. Waters 2008 hints at a similar idea.

18. *Xenopus laevis* is the African clawed frog, which became an important model organism for vertebrate development. One of its main advantages is that it breeds all year round, being a tropical species. Spemann and colleagues could only do experiments with their northern newts in spring.

19. The extent to which the world is neatly divided into levels has been called into question. (For a challenging discussion, see Potochnik 2017.) I use the term "level" here mainly to refer to the domain of objects that the different sciences study (by their own lights). CEE studies embryonic tissues and

cells; molecular biology studies molecules. Inter-level practices intervene on both kinds of objects. Those who are skeptical of levels could think of them simply as domain-crossing practices.

20. Inter-level investigative practices in my sense do not necessarily involve inter-level experiments in the sense of Craver 2007. In such an experiment, an entity at some level is used to manipulate an entity at a different level. My inter-level practices do not require such inter-level interventions, although I don't want to rule them out. All they require is that interventions at different levels are part of the same practice, where practices are individuated by their goals (e.g., cloning genes involved in development).

21. I tend to think that there was less consensus than in Kuhn's influential image of "normal science," but this would have to be investigated more closely.

22. See Minelli and Pradeu 2014, in particular the contributions by Thomas Pradeu and Alan Love for differing views about this topic.

REFERENCES

Arias, A. M., and B. Steventon. 2018. "On the Nature and Function of Organizers." *Development* 145: 5.

Bechtel, W. 2006. *Discovering Cell Mechanisms: The Creation of Modern Cell Biology*. Cambridge: Cambridge University Press.

Blumberg, B., C. V. Wright, E. M. De Robertis, and K. W. Cho. 1991. "Organizer-Specific Homeobox Genes in *Xenopus laevis* Embryos." *Science* 253, no. 5016: 194–96.

Bourrat, P. 2019. "On Calcott's Permissive and Instructive Cause Distinction." *Biology and Philosophy* 34: 1.

Bridgman, P. W. 1927. *The Logic of Modern Physics*. New York: MacMillan.

Calcott, B. 2017. "Causal Specificity and the Instructive–Permissive Distinction." *Biology and Philosophy* 32: 481–505.

Chang, H. 2007. *Inventing Temperature: Measurement and Scientific Progress*. Oxford: Oxford University Press.

Cho, K. W. Y., B. Blumberg, H. Steinbeisser, and E. M. De Robertis. 1991. "Molecular Nature of Spemann's Organizer: The Role of the *Xenopus* Homeobox Gene Goosecoid." *Cell* 67: 1111–20.

Craver, C. 2007. *Explaining the Brain: Mechanisms and the Mosaic Unity of Neuroscience*. Oxford: Oxford University Press.

De Robertis, E. M., M. Blum, C. Niehrs, and H. Steinbeisser. 1992. "Goosecoid and the Organizer." *Development* 116: 167–71.

De Robertis, E. M. 2006. "Spemann's Organizer and Self-Regulation in Amphibian Embryos." *Nature Reviews. Molecular Cell Biology* 7: 296–302.

Dizadji-Bahmani, F., R. Frigg, and S. Hartmann. 2010. "Who's Afraid of Nagelian Reduction?" *Erkenntnis* 73: 393–412.

Dobzhansky, T. 1937. *Genetics and the Origin of Species.* New York: Columbia University Press.

Driesch, H. 1892. "Entwicklungsmechanische Studien. I. Der Wert der beiden ersten Furchungszellen in der Echinodermen-Entwicklung. Experimentelle Erzeugung von Theil- und Doppelbildungen." *Zeitschrift für Wissenschaftliche Zoologie* 53: 160–78.

Feest, U. 2005. "Operationism in Psychology—What the Debate Is About, What the Debate Should Be About." *Journal for the History of the Behavioral Sciences* 41: 131–50.

Feest, U. 2010. "Concepts as Tools in the Experimental Generation of Knowledge in Cognitive Neuropsychology." *Spontaneous Generations* 4: 173–90.

Gehring, W. J. 1998. *Master Control Genes in Development and Evolution: The Homeobox Story.* New Haven, Conn.: Yale University Press.

Gilbert, S. F., and M. J. F. Barresi. 2016. *Developmental Biology.* 11th ed. Sunderland, Mass.: Sinauer Associates.

Grainger, R. M. 1992. "Embryonic Lens Induction: Shedding Light on Vertebrate Tissue Determination." *Trends in Genetics* 8, no. 10: 349–55.

Griffiths, P. E., A. Pocheville, B. Calcott, K. Stotz, H. Kim, and R. Knight. 2015. "Measuring Causal Specificity." *Philosophy of Science* 82: 529–55.

Halder, G., P. Callaerts, and W. J. Gehring. 1995. "Induction of Ectopic Eyes by Targeted Expression of the Eyeless Gene in *Drosophila*." *Science* 267: 1788–92.

Hamburger, V. 1988. *The Heritage of Experimental Embryology. Hans Spemann and the Organizer.* Oxford: Oxford University Press.

Hoyningen-Huene, P. 1997. "Comment on J. Kim's 'Supervenience, Emergence, and Realization in the Philosophy of Mind.'" In *Mindscapes: Philosophy, Science, and the Mind,* edited by M. Carrier and P. K. Machamer, 294–302. Konstanz/Pittsburgh: Universitätsverlag Konstanz/University of Pittsburgh Press.

Hull, D. 1974. *Philosophy of Biological Science.* Englewood Cliffs, N.J.: Prentice Hall.

Kaiser, M. I. 2015. *Reductive Explanation in the Biological Sciences.* Berlin: Springer International Publishing. https://www.springer.com/de/book/9783319253084.

Kim, J. 2007. *Physicalism, or Something Near Enough*. Princeton, N.J.: Princeton University Press.

Kitcher, P. 1984. "1953 and All That. A Tale of Two Sciences." *Philosophical Review* 93: 335–73.

Konopka, R. J., and S. Benzer. 1971. "Clock Mutants of *Drosophila melanogaster.*" *Proceedings of the National Academy of Sciences* 68: 2112–16.

Kripke, S. A. 1980. *Naming and Necessity*. Cambridge, Mass.: Harvard University Press.

Kuhn, T. S. 1970. *The Structure of Scientific Revolutions,* 2nd ed. Chicago: University of Chicago Press.

Lewis, W. H. 1904. "Experimental Studies on the Development of the Eye in Amphibia: I. On the Origin of the Lens, Rana Palustris." *American Journal of Anatomy* 3: 505–36.

Maienschein, J. 1991. "The Origins of Entwicklungsmechanik." In *A Conceptual History of Modern Embryology*, 43–61. London: Plenum Press.

Maniatis, T., S. G. Kee, A. Efstratiadis, and F. C. Kafatos. 1976. "Amplification and Characterization of a Beta-Globin Gene Synthesized in Vitro." *Cell* 8: 163–82.

Minelli, A., and T. Pradeu, eds. 2014. *Towards a Theory of Development*. Oxford: Oxford University Press.

Nagel, E. 1961. *The Structure of Science. Problems in the Logic of Scientific Explanation*. London: Routledge and Kegan Paul.

Nickles, T. 1973. "Two Concepts of Intertheoretic Reduction." *The Journal of Philosophy* 70: 181–220.

Nurse, P. 1975. "Genetic Control of Cell Size at Cell Division in Yeast." *Nature* 256: 547.

Ogino, H., and K. Yasuda. 2000. "Sequential Activation of Transcription Factors in Lens Induction." *Development, Growth & Differentiation* 42: 437–48.

Ogino, H., H. Ochi, H. M. Reza, and K. Yasuda. 2012. "Transcription Factors Involved in Lens Development from the Preplacodal Ectoderm." *Developmental Biology* 363: 333–47.

Oyama, S. 2000. *The Ontogeny of Information: Developmental Systems and Evolution*, 2nd ed. Durham, N.C>: Duke University Press.

Potochnik, A. 2017. *Idealization and the Aims of Science*, 1st ed. Chicago: University of Chicago Press.

Putnam, H. 1975. "The Meaning of 'Meaning.'" In *Language, Mind and Knowledge, Vol. VII*, edited by K. Gunderson, 131–93. Minneapolis: University of Minnesota Press.

Rosenberg, A. 1985. *The Structure of Biological Science.* Cambridge: Cambridge University Press.

Rosenberg, A. 1997. "Reductionism Redux: Computing the Embryo." *Biology and Philosophy* 12: 445–70.

Rosenberg, A. 2006. *Darwinian Reductionism.* Chicago: University of Chicago Press.

Roux, W. 1888. "Beiträge Zur Entwickelungsmechanik des Embryo. Über die künstliche Hervorbringung halber Embryonen durch Zerstörung einer der beiden ersten Furchungskugeln, sowie über die Nachentwicklung (Postgeneration) der fehlenden Körperhälfte." *Virchow's Archiv Für Pathologische Anatomie Und Physiologie Und Für Klinische Medizin* 114: 113–53.

Saha, M. 1991. "Spemann Seen through a Lens." In *A Conceptual History of Modern Embryology,* edited by S. F. Gilbert, 91–108. New York: Plenum Press.

Saha, M. S., C. L. Spann, and R. M. Grainger. 1989. "Embryonic Lens Induction: More Than Meets the Optic Vesicle." *Cell Differentiation and Development: The Official Journal of the International Society of Developmental Biologists* 28: 153–71.

Schaffner, K. F. 1969. "The Watson-Crick Model and Reductionism." *British Journal for the Philosophy of Science* 20: 325–48.

Smith, W. C., and R. M. Harland. 1992. "Expression Cloning of Noggin, a New Dorsalizing Factor Localized to the Spemann Organizer in *Xenopus* Embryos." *Cell* 70: 829–40.

Spemann, H. 1901. "Über Correlationen in Der Entwicklung Des Auges." *Verhandlungen Der Anatomischen Gesellschaft* 15: 61–79.

Spemann, H. 1936. *Experimentelle Beiträge Zu Einer Theorie Der Entwicklung.* Berlin: Julius Springer.

Vance, R. E. 1996. "Heroic Antireductionism and Genetics: A Tale of One Science." *Philosophy of Science* 63: S36–45.

Waters, C. K. 2007. "Causes That Make a Difference." *The Journal of Philosophy* 104, no. 11: 551–79.

Waters, C. K. 2008. "Beyond Theoretical Reduction and Layer-Cake Antireduction: How DNA Retooled Genetics and Transformed Biological Practice." *The Oxford Handbook of Philosophy of Biology.* http://www.oxfordhandbooks.com/view/10.1093/oxfordhb/9780195182057.001.0001/oxf.

Waters, C. K. 2017. "No General Structure." In *Metaphysics and the Philosophy of Science: New Essays,* edited by M. H. Slater and Z. Yudell, 81–108. Oxford: Oxford University Press.

Weber, M. 1999. "Hans Drieschs Argumente für den Vitalismus." *Philosophia Naturalis* 36: 265–95.

Weber, M. 2004. "Walking on the Chromosome: Drosophila and the Molecularization of Development." In *From Molecular Genetics to Genomics: The Mapping Cultures of Twentieth-Century Genetics*, edited by Jean-Paul Gaudillière and Hans-Jörg Rheinberger, 63–78. London: Routledge.

Weber, M. 2005. *Philosophy of Experimental Biology*. Cambridge: Cambridge University Press.

Weber, M. 2006. "The Central Dogma as a Thesis of Causal Specificity." *History and Philosophy of the Life Sciences* 28: 565–80.

Weber, M. 2017. "Which Kind of Causal Specificity Matters Biologically?" *Philosophy of Science* 84: 574–85.

Weber, M. 2022. *Philosophy of Developmental Biology*. Cambridge: Cambridge University Press.

Weber, M. Forthcoming. "Causal Selection versus Causal Parity in Biology: Relevant Counterfactuals and Biologically Normal Interventions." In *Philosophical Perspectives on Causal Reasoning in Biology*, edited by B. J. Hanley and C. K. Waters. Minneapolis: University of Minnesota Press.

Weinberg, S. 1992. *Dreams of a Final Theory*. New York: Pantheon.

Wimsatt, W. C. 1976. "Reductive Explanation: A Functional Account." In *PSA 1974 (Proceedings of the 1974 Biennial Meeting, Philosophy of Science Association), Vol. 32*, edited by R. S. Cohen, C. A. Hooker, A. C. Michalos, and J. W. van Evra, 671–710. Dordrecht/Boston: Reidel.

Woodward, J. 2003. *Making Things Happen: A Theory of Causal Explanation*. New York: Oxford University Press.

Woodward, J. 2010. "Causation in Biology: Stability, Specificity, and the Choice of Levels of Explanation." *Biology and Philosophy* 25: 287–318.

8

EXPLANATION IN CONTEXTS OF CAUSAL COMPLEXITY
Lessons from Psychiatric Genetics

LAUREN N. ROSS

Some have claimed that psychiatry is in a "crisis" (Hyman 2013; Morgan 2015; Poland and Tekin 2017). These claims often target the lack of known or identifiable causal etiologies for psychiatric diseases, suggesting that they are "among the most intractable enigmas in medicine" (Sullivan, Daly, and O'Donovan 2012, 537). While the intractable nature of these disorders is often associated with their "causal complexity" (Poland and Tekin 2017, 5), it is not always clear exactly what is meant by this. How should we understand causal complexity in this domain? How does it challenge scientific efforts to understand and explain these diseases? This chapter addresses these questions by examining two main types of causal complexity in psychiatry. My analysis clarifies what these types of causal complexity are, how they challenge efforts to understand and explain these disorders, and how scientists are working to overcome these challenges.

1. INTRODUCTION

Over the past decade, there have been increasingly common claims that psychiatry is in a "crisis" (Hyman 2013; Morgan 2015; Poland and Tekin 2017)—that it is an "embryonic" and "immature" science that remains in its "early stages" (Hyman 2010, 155, 171; 2013). According to these views, psychiatry is stuck within a disease framework that is "seriously flawed" (Poland and Tekin 2017, 1) and marked by "incredible insecurity" and "nosologic instability" that are "beyond a full resolution" (Kendler and Zachar 2008, 370–71). Many of these criticisms target the lack of known or identifiable causal etiologies for psychiatric disorders. This, of course, is compared to the relative success that has been enjoyed in identifying such etiologies for various nonpsychiatric or "physical medicine" diseases. It has been suggested that "psychiatric disorders are among the most intractable enigmas in medicine" and that they "have been intractable to approaches that were fruitful in other areas" of medical science (Sullivan, Daly, and O'Donovan 2012, 537). The intractable nature of these disorders is often associated with their "causal complexity" (Poland and Tekin 2017, 5), where this is interpreted in a variety of ways. On one interpretation, causal complexity is connected with views that the human brain is "the most complex object in the known universe" due to its large number of neurons and synaptic connections (Hoffecker 2011, ix). A second interpretation suggests that psychiatric disorders are complicated at the level of etiology or in terms of the causal processes that produce them (Uher and Zwicker 2017). A third interpretation suggests that the genetic bases and heritability of mental disorders are complex in ways that we might not see with other conditions (Lemoine 2016; Tsuang, Glatt, and Faraone 2006; Mitchell 2012).

Despite efforts to provide clarity, it is not always clear exactly what is meant by "causal complexity" and how it leads to the "intractable" nature of these disorders. These points raise a number of questions. First, how should we understand causal complexity in this domain? Second, if causal complexity makes sense of the "intractable" and "enigmatic" nature of psychiatric disease, how exactly does it challenge our scientific efforts to understand and explain it?

This chapter addresses these questions by analyzing two types of causal complexity that are common in psychiatry and that challenge efforts to understand and explain these disorders. My analysis clarifies what these types of causal complexity are, how they challenge efforts to understand and explain psychiatric disease, and how scientists are working to overcome these challenges. This analysis examines work in psychiatric genetics where genome-wide association studies (GWAS) have been used to search for genetic causes of disease. I do not claim that genetic factors are the only relevant (or even the main) causes of these diseases. Instead, I suggest that examining scientific efforts to identify such causes reveals important types of causal complexity that emerge in this domain. As will become clear, one main suggestion of this analysis is that while these types of complexity are particularly common and troubling in psychiatry, they are actually found throughout many areas of medicine. The rest of this chapter is structured as follows. In section 2, I provide some background on disease causation, including particular causal standards that ideal diseases are often expected to meet. In sections 3 and 4, I examine two different types of causal complexity, which I refer to as multicausality and causal heterogeneity. These sections discuss how these types of causal complexity should be understood, how they challenge disease explanation, and how scientists are working to overcome these challenges. In section 5, I examine a further challenge for disease explanation that is relatively unique to biomedicine and that has received little to no attention in the philosophical literature. Section 6 concludes.

2. A CAUSAL FRAMEWORK FOR DISEASE: SOME BACKGROUND

At a basic level, disease explanation involves a disease phenotype (D) and its causes or causal etiology (C). This setup helps clarify a common two-step process for discovering new diseases that has been employed from Hippocratic to modern times. In the first step, a disease phenotype (D) is associated with some symptomology that recurs, with variation, across patients.[1] A second step in this process involves identifying the causal factors (C) or the causal etiology that produces this disease phenotype.[2] While various "physical" or "somatic" diseases have known causal etiologies, most if not all psychiatric conditions are of unknown etiology. In this sense, most psychiatric disorders are stuck at this first stage of discovery. Researchers have identified the symptomology that they think characterizes these conditions, but they do not yet know what causes them. This causal information is essential for ensuring that a disease category is valid—it guides how researchers and physicians classify, explain, and discover "bona fide" disease traits (Hyman 2010). Identifying etiology is valuable because it can be targeted to explain, predict, and control disease occurrence. While symptomology can suggest palliative treatments that comfort and mask symptoms, it usually cannot suggest curative measures or inform disease explanation, as both require targeting the root cause of disease.

In this sense, causal etiology serves as a gold standard for many interrelated projects in medicine, including disease classification, explanation, discovery, and treatment. Unsurprisingly, psychiatric conditions can face significant scrutiny when their etiologies are unknown. In particular, if the causal etiology of a purported psychiatric disease is unknown, the

"legitimacy" and "validity" of the disease are often questioned. This is captured by the modern medical view that "if you cannot explain a distinct
and unambiguous etiology for a syndrome, preferably in biological terms,
then you do not have a *real* disorder" (Kendler 2012, 1, emphasis original).
This is not to say that the medical community questions whether patients
actually experience these symptoms. Instead, they question whether the
disease category associated with these symptoms will remain stable and un-
changed as more is uncovered about its causal etiology (Kendler and Zachar
2008). Why this worry? One lesson that diseases have repeatedly taught us
is that symptomology is a rough and unpredictable guide to causal etiology
(Hyman 2010, 161). The repeated presentation of clear-cut symptom clusters across patients is no guarantee that these symptoms all arise from the
same causal process. We see this in cases where the same etiology produces
different symptoms and where different etiologies produce the same symptoms (Ross forthcoming).[3]

A main goal of psychiatry is to get to this second step of disease discovery and identify the causal etiologies of these conditions (Sullivan, Daly, and
O'Donovan 2012, 537). One strategy that is used to achieve this goal involves
collecting patients with the same diagnosis and searching among them for
the factors that they have in common and that might be causally responsible for their disease. This involves starting with some phenotype of interest
(D) and then searching backward or causally upstream to identify its causes
(C).[4] This basic strategy has been implemented in genome-wide association
studies (GWAS). These studies analyze the genomes of patients with particular psychiatric disorders in order to identify those gene variants that
they all share and that potentially cause these diseases. Expectations about
the type of results these studies should provide have been influenced by an
"ideal" model of disease causation that continues to figure in modern medicine. This ideal model—sometimes referred to as the "hard" medical model
or the "biomedical" model (Kendler 2012; Engel 1977)—originated with
nineteenth-century germ theory and contains two main causal standards
(Ross 2018). First, this model involves a (1a) single cause standard, which
maintains that a particular instance of some disease has one main causal
factor. Second, this model also involves a (2a) shared cause standard, which
maintains that all instances of a particular disease have the same (or some
similar) causal process. This model captures the expectation that diseases
should have single, shared causal etiologies.

Although some diseases meet the strict standards captured in this "ideal" model, most do not. GWAS have provided further evidence for the claim that psychiatric disorders often fail to fit this model. In particular, these studies have identified two types of causal complexity that capture ways in which this ideal model breaks down. First, these studies indicate that some psychiatric disorders are characterized by (1b) *multicausality* in the sense that each instance of the disease is caused by many gene variants that work together in aggregate to produce the condition. This finding conflicts with the single cause standard or monocausal-type picture. Second, these results also suggest that some psychiatric disorders are (2b) *causally heterogeneous* in the sense that distinct instances of the same disease are caused by different combinations of gene variants. This conflicts with the shared cause standard, as different combinations of causes are capable of producing the same disease.

This breakdown provides a helpful way to understand four distinct causal architectures (1a, 1b, 2a, 2b) and two types of causal complexity—(1b) multicausality and (2b) causal heterogeneity—as outlined in Figure 8.1. In this figure, each causal architecture has to do with how "simple" or "complex" causal factors are with respect to some specified effect of interest.[5] This figure shows how each of these four architectures are related to each other and how they come apart. As monocausality and multicausality have to do with the number of causes for a single instance of disease, they operate at the token level. They represent two sides of the spectrum for token causal etiology—one more complex (1b) and the other less so (1a). As causal homogeneity and causal heterogeneity have to do with whether causes are similar or different across cases of disease, they operate at the type or population level. These also represent two sides of a spectrum, but in this case for type causal etiology—one more complex (2b) and the other less so (2a). These token- (1a, 1b) and type- (2a, 2b) level causal architectures are not mutually exclusive. Knowing that a type-level disease trait is causally heterogeneous or homogeneous provides no information about whether its instances are multicausal or monocausal, and vice versa.[6] The category that a disease falls into on the left side of Figure 8.1 does not dictate or influence which category it falls into on the right side (and vice versa). Diseases that meet the less complex causal architectures (1a, 2a) come with particular advantages, while diseases that meet the more complex ones (1b, 2b) involve various challenges for understanding, explanation, classification, and control. I discuss these

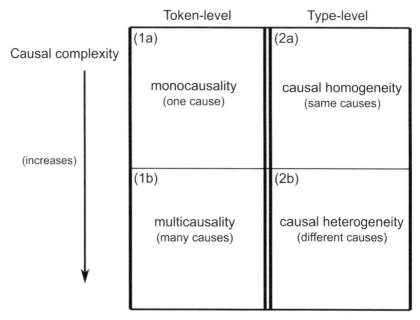

Figure 8.1. Four different causal architectures (1a, 2a, 1b, 2b) and two different types of causal complexity (1b, 2b).

types of causal complexity in more detail, the various challenges that they present, and how scientists work to overcome these challenges.

3. MULTICAUSALITY

As briefly described previously, multicausality can be thought of as contrasting with monocausality or the well-known "monocausal model" of disease. The former involves a disease instance that has many causes, while the latter involves a disease instance that has one main cause. Various conditions are thought to fit this monocausal picture, such as scurvy, tuberculosis, chicken pox, giardiasis, and Huntington's disease, among others. Genetic conditions that fit this monocausal model are often referred to as "single-gene," "monogenic," or "Mendelian" diseases as opposed to diseases that are "polygenic" or "complex" (Cooper et al. 2013; Kendler 2005; Torkamani, Wineinger, and Topol 2018; Mitchell 2012). What does it mean to say that these diseases each have single main causes? How could any disease have a single main cause? Addressing these questions requires specifying what is meant by "causation" and how one factor could be privileged as the main or most important cause of some outcome. In this chapter, I rely on an interventionist account of causation, in which a causal factor "makes a difference" to its effect in the sense of providing control over it (Woodward 2003). On this account, to say that C is a cause of D means that an intervention that changes the values of C and no other variables in background circumstances B produces changes in the values of D. In other words, causes are factors that operate like handles or switches in the sense that they can be potentially manipulated to provide control over their effects. Manipulating these factors produces changes in the effects they are related to. Importantly, this account does not require that such an intervention is currently or technologically available, but just that *if* such an intervention were performed, the ensuing change in the effect variable *would* occur (Woodward 2003, 11). Notice that for the diseases mentioned previously, each has a particular factor such that if that factor were manipulated, it would control the occurrence and nonoccurrence of the disease in question. For example, manipulating dietary vitamin C provides control over whether a patient acquires scurvy. The same could be said for interventions on the single main causes of the other monocausal diseases mentioned previously.[7] Manipulating these factors provides control over the occurrence and nonoccurrence of these disorders—these causes are targeted in treating, preventing, explaining, and controlling these diseases.

These monocausal diseases have another important feature. The particular factors that are identified as the single main causes of these diseases have a special type of control over them.[8] These factors have probable control over disease traits in the sense that manipulating these causes provides a high probability of producing the occurrence and nonoccurrence of the trait.[9] In order to see this, consider a light switch on a wall and the different degrees of probable control it can exhibit over the state of the light being "on" or "off." In a first system, flipping a switch up provides a 60 percent chance that the light turns "on," while flipping it down provides a 60 percent chance that the light turns "off." In a second system, flipping the switch up provides nearly a 100 percent chance that the light turns "on," while flipping it down provides nearly a 100 percent chance of the light turning "off." The switch in the second system has a higher degree of probable control over the light than the first because manipulating this switch provides a higher likelihood of changing (or controlling) the state of the light. Paradigmatic monocausal diseases approximate the second switch system. Factors that are identified as the single main causes of these diseases provide a high degree of probable control over them. This can be seen in Huntington's disease, which is caused by a mutation in the *huntingtin* gene. When a patient has this mutation, her likelihood of acquiring the disease is nearly 100 percent, and if she lacks the mutation, her likelihood of *not* acquiring it is nearly 100 percent. Identifying single causes with a high degree of probable control is very valuable in medicine. These factors provide a reliable indication of whether a disease will manifest or not and they identify factors that can be targeted to control, explain, treat, and prevent disease.

Determining whether a *single* factor has probable control over a disease involves assessing potential background conditions that may also influence the disease outcome. Paradigmatic monocausal diseases have single causes with probable control, where this control is stable across changes in common or relevant background conditions. For example, as Kendler states, "If you have one copy of the pathogenic gene for Huntington's disease, it does not matter what your diet is, whether your parents were loving or harsh, or if your peer group in adolescence were boy scouts or petty criminals. If you have the mutated gene and you live long enough, you will develop the disease" (Kendler 2005, 394). In other words, there are no additional genetic, environmental, or other factors that influence or alter the cause–effect relationship in question (Kendler 2005, 397).

The stability of this probable control is related to the genetic concepts of penetrance and effect size.[10] Penetrance refers to the percentage of individuals in a population with a particular genotype who exhibit the corresponding phenotype, where phenotype is either present or absent. If a gene variant is 30 percent penetrant, then 30 percent of those individuals with the genotype will express the phenotype. Alternatively, the variant that causes HD (Huntington's disease) is 100 percent penetrant—or "fully" and "completely" penetrant—because 100 percent of those individuals with this gene variant will express this disease phenotype (Stewart et al. 2007). This measure can be thought of as giving an indication of a gene's ability to "penetrate through to the phenotype" despite changes in other background factors (Carr 2014). In this sense, complete or high penetrance refers to "determinative" genes for which "environmental and other factors have little effect on the phenotype" (Weiss 2007; Carr 2014, 283). Probable control is also related to the genetic concept of "effect size," which concerns the proportion of variation in the phenotype that is "explained by" or "attributed to" variation in the genotype (Nakagawa and Cuthill 2007; Maier et al. 2017). Effect size is often described as capturing the "magnitude of an effect" that genotype has over phenotype, and it is often used synonymously with the notion of "heritabil-ity" (Nakagawa and Cuthill 2007, 593; Maier et al. 2017). Genes that are con-sidered the single main causal factor for a phenotype often have large effect sizes. In these cases, variation in the population-wide phenotype is explained by variation in a gene.

In the early stages of GWAS, researchers were hoping to identify gene variants with a high degree of probable causal control, a high degree of pene-trance, and large effect sizes. Instead, GWAS uncovered nearly the opposite type of finding. These studies identified gene variants with a low degree of probable control, little penetrance, and small effect sizes. In other words, they identified genes that were "packing much less of a phenotypic punch than expected" (Goldstein 2009, 1696). When researchers find single gene variants with low probable control, variable penetrance, or small effect sizes, they of-ten interpret this as an indication that other causal factors—such as other genes, environmental variables, and so on—influence and interact with these variants in producing the disease (Griffiths et al. 2008, 249). This is to say that they view these diseases as multicausal—as produced by many causal factors that work together in aggregate to produce the condition.[11] Instead of having a single gene that fully "penetrates" through to the disease trait, these genes depend on and interact with other causes in producing the trait. As Cooper states, "most carriers of the risk alleles discovered by genome-wide associa-tion studies (GWAS) may never develop the disease in question . . . because these variants generally only make a small contribution to the multifactorial aetiology of the condition" (Cooper et al. 2013, 1078). Researchers expect disease etiology to include factors with a high degree of probable control over disease. When a single causal factor fails to provide this type of control, they search for multiple factors that provide this type of control together. In these cases, the disease is considered multicausal as it is produced by many causes that all have a "collective impact" on the disease outcome (Ideker, Dutkowski, and Hood 2011, 3). A simple example of this is phenylketonuria (PKU), which is caused by two main factors: a gene variant and a dietary fac-tor. Acquiring this disease requires the presence of both of these factors, and both are required to gain probable control over the disease state (Murphy 1997, 113). Manipulating the variant only provides control over the disease when the dietary factor is present, and manipulating the dietary factor only provides control when the gene is present. In this sense, PKU is multicausal because it takes more than one causal factor to gain probable control over the occurrence and nonoccurrence of this disease.[12]

This clarifies the rationale behind identifying a disease as monocausal versus multicausal. The number of relevant causal factors is determined on the basis of the number it takes to achieve a high degree of probable causal control over the disease trait. In this sense, the results of GWAS further support the view that most psychiatric diseases are multicausal in etiology (Price, Spencer, and Donnelly 2015). As Plomin and Kovas (2005, 600) state, "it is now generally accepted that genetic influence on common disorders is caused by multiple genes of small effect size rather than a single gene of major effect size." Researchers do not deny that monocausal diseases exist, they just think that most of them have already been identified. In other words, diseases with single genetic causes that have "large effect sizes—the low-hanging fruit—have already been detected" (Park et al. 2010, 570). What we have left are more complicated multicausal diseases that are much more challenging to discover and understand. As Goldstein states, "the modest size of genetic effects detected so far confirms the multifactorial aetiology of these conditions and suggests that complex diseases will require substantially greater research effort to detect additional genetic influences" (Goldstein 2009, 9).

How exactly does multicausality challenge scientific efforts to understand and explain these diseases? A first challenge with this type of causal complexity is that it requires identifying the many causal factors involved in producing an effect. Where providing a causal explanation of some effect involves identifying and citing its causes, this becomes more and more difficult as the number of relevant causes increases. More explanatorily relevant causal factors mean more factors to identify and appeal to. Furthermore, most psychiatric diseases do not appear to be similar to PKU in the sense of having two main causes. Researchers hypothesize that some psychiatric diseases have hundreds of causally relevant gene variants, representing a far more extreme case of multicausality than PKU. Second, it is not enough to simply identify these factors—these explanations require providing some coherent story about how these factors work together to produce the disease in question. This includes specifying how various factors depend on each other in producing disease, what role they play in the pathogenic process, and what their particular effect sizes are. A third main challenge with this type of causal complexity is that scientists appear to be unsatisfied by explanations that are *too* multicausal. In cases where multicausal genetic factors balloon out to an extreme degree, scientists often suggest that these factors fail to capture the right "level" of "causal action" for the disease (Kendler

2013, 1060). When this happens in the context of genetics, researchers can claim that such causes provide little guidance or understanding and that there is likely some alternative "level" that better captures the relevant causal etiology. This is mentioned by Goldstein, who states that "if effect sizes were so small as to require a large chunk of the genome to explain the genetic component of a disorder, then no guidance would be provided: in pointing at everything, genetics would point at nothing" (2009, 1696). This third challenge is that multicausality in the extreme is inadequate for explanation and that it suggests that the causally relevant factors are likely found at some "level" other than molecular biology. This is driven by the expectation that the right level or characterization of causal etiology should be somewhat unified and not too splintered.

If multicausality poses these challenges to understanding and explanation, how do scientists overcome them? First, the challenge of identifying many causal factors has been approached, in part, by modifying search methods such that they are better equipped to identify causes with small effect sizes (Park et al. 2010). The second challenge—providing a coherent story about how these factors work together—is addressed by various strategies aimed at unification. Here the unification is focused on *interaction* and specifying how causes are unified on the basis of all interacting together in a single causal process that produces disease. This is sometimes accomplished by providing a "unifying" mechanism or pathway that integrates all causes with respect to the effect of interest. These unifying causal processes can clarify how the many causes interact with each other, the step-by-step or sequential order of their operation, and the magnitude of their individual effects over the outcome of interest. For example, this can be done by identifying "multiple related genes in the same functional pathway" that "work together to confer disease susceptibility" (Wang, Li, and Hakonarson 2010). When gene variants are unified in this way, it can allow for the identification of single, unified causal processes at higher "levels." For example, multiple gene variants may all influence a higher-level cellular process, where this process captures how they all interact to produce disease. This move to a higher level can circumvent the issue of rampant multicausality at the level of gene variants. Instead of appealing to many lower-level splintered causes, this provides the option of citing a single, unified, and coherent higher-level causal process. This strategy of unification can be understood as reworking or converting a situation of "many" causes into a situation in which "one" cause or causal process is responsible. This converts "many" causes into "one" causal process, in which this process is typically a single mechanism or pathway. In the context of explaining a particular effect, this suggests that there is something useful about ascribing causal responsibility to a single, causal entity as opposed to pointing to some distributed set of seemingly unrelated factors.

4. CAUSAL HETEROGENEITY

A second type of causal complexity in this domain is causal heterogeneity, which contrasts with causal homogeneity. Starting with the latter, causal homogeneity refers to a situation in which distinct instances of the same effect (in this case, a disease trait) are produced by the same combination of causes. In other words, these causes are "homogeneous" across different instances of the same type of effect. Many diseases that fit the monocausal model provide straightforward examples of this causal architecture. This can be seen in the case of scurvy because every instance of this disease is caused by the same factor (namely, a deficiency of dietary vitamin C). Causal homogeneity can also be met by multicausal diseases, so long as every instance of the disease is produced by the same combination of causes. An example of this is PKU because every instance of this disease is caused by the same two factors. This shows how causal homogeneity is distinct from monocausality and multicausality. While monocausality and multicausality have to do with the number of causes for each instance of an effect, causal homogeneity has to do with whether these causes are similar or different across all of these instances.

In modern medicine, there are particular assumptions about disease causation that involve causal homogeneity. In particular, there is a common default assumption that in order for a disease trait to be "valid" and "legitimate" it should be causally homogeneous in the sense of having some shared causal etiology (Ross 2018). This notion of shared etiology is sometimes referred to as a "disorder-specific pathophysiology" (Caspi and Moffitt 2006, 586), a "shared causal process" (Zachar 2014, 87), a "shared pathogenesis," or the "causal signature" for a particular disease (Murphy 2006, 105). As these shared causes capture what unifies various instances under the same disease heading, they are referred to as "unifying causes" or the "unifying theoretical underpinning" for a given disease (Egger 2012, 1). In current medical theory, it is often expected that diseases have some unifying and singular causal story—that they have "single biological essences" at the level of etiology (Kendler 2012, 1). The presence of this assumption about shared causal etiology is seen in various medical contexts. It figures in decisions about what are deemed "valid" disease traits and how such traits and their etiologies should be discovered. For example, in the context of psychiatry, "diagnostic validity" is defined in terms of shared causal etiology. In particular, "diagnostic validity" is "shorthand to signify definitions that capture families of closely related disorders with similar pathophysiology" (Hyman 2010, 162). Additionally, this assumption figures in GWAS and other studies that aim to discover disease and disease etiology. This is because such studies group together patients with similar symptomology in the hope of finding some causal process that they all share or have in common. As Maier and colleagues state, "most genetic studies are based on the assumption that individuals who exhibit similar symptoms or who have been diagnosed with the same disease are representatives of the *same* underlying biology defined by a *common* genetic architecture" (2017, 1063, emphasis added). These strategies assume that distinct instances of the same disease are all produced by similar or homogeneous causes.

Although causal homogeneity has figured into the setup and expectations of GWAS, many purported disease traits have failed to meet this standard. Emerging results suggest that some psychiatric diseases are causally heterogeneous—or exhibit "etiological heterogeneity"—in the sense that distinct instances of the same disease are caused by different combinations of causal factors.[13] Psychiatric disorders that are thought to exhibit this type of causal complexity include schizophrenia, autism spectrum disorder, and bipolar disorder (Betancur 2011; Takahashi 2013). Causal heterogeneity is also present in other nonpsychiatric (or physical medicine) diseases.[14] An example of this causal architecture is seen in Parkinson's disease, which can be produced by different causal factors in different patients with this same disease. This disease can be produced by single gene variants (C_1), single environmental factors (C_2), and combinations of genetic and environmental factors (C_3) (Nandipati and Litvan 2016). In cases where the heterogeneous causes are genetic, the disease is referred to as "genetically heterogeneous" (Barondes 1992, 299). In this sense, "genetic heterogeneity can be defined as mutations at two or more genetic loci that produce the same or similar phenotypes (either biochemical or clinical)" (McGinniss and Kaback 2013, 7).[15] An example of this is retinitis pigmentosa, which can be caused by anywhere from 75 to 300 different gene mutations that can each "act alone" to produce the disease (Hyman 2010, 163).

A key feature of causal heterogeneity is that it involves a many-to-one relationship between disease causes and the disease effect. In the context of genetically heterogeneous traits, this results in a "phenotypic convergence of independent mutations" that can involve "diverse genetic pathways to similar disease traits" or to some "common symptomology" (Hyman 2010, 163; Takahashi 2013, 648). This is a kind of funneling of different causal factors or pathways onto the same final effect of interest. The causal starting points of this funnel are often described as each individually sufficient or able to "act alone" in producing the final disease outcome (Hyman 2010, 163). In the context of genetics, this ability to act alone is captured by the fact that each heterogeneous variant is "highly penetrant" for the disease. We see this in the case of retinitis pigmentosa, in which "each deleterious mutation acts as a single gene 'Mendelian' disorder within a family, but in aggregate, different families are affected by a large number of distinct mutations in different genes" (Hyman 2010, 163). In these cases, the heterogeneous genetic causes are sometimes referred to as "rare variants" because the etiology can be so varied or heterogeneous that any causal variant only occurs very "rarely," sometimes only in those individuals of a single family. From the standpoint of any particular heterogeneous gene variant, each can have a high probability of producing the disease. However, from the standpoint of the population-wide disease trait, there is no single genetic cause that is responsible for all instances of the disease. This shows up in the fact that these gene variants have small effect sizes. Variation in any individual gene variant only explains a small percentage of the variation in the population-wide trait. Some of this variation is explained by the other gene variants that are also capable of producing the disease.

How does causal heterogeneity challenge efforts to understand and explain scientific phenomena? A first challenge is that heterogeneous causes are limited in providing explanations of type-level phenomena because no single heterogeneous cause explains all instances of its type-level effect. Heterogeneous causes are explanatorily and causally relevant to a fraction of all instances of their effect as opposed to having this type of relevance to most or all of these instances. In order to see this, consider the case of Parkinson's disease, which has three different individually sufficient causes (C_1, C_2, and C_3). Appealing to any one of these causes (e.g., C_1) would fail to provide an adequate explanation of the population-wide disease trait because no heterogeneous cause alone "makes a difference" to *all* instances of this trait. Similarly, targeting one of these factors will fail to provide causal control over most or all cases of this disease.[16] This, of course, is because some instances of the disease are caused by different causal factors entirely (e.g., C_2 and C_3).[17] Thus, one problem for heterogeneous causes is that they have causal relevance and control of "narrow scope" over the type-level disease trait. Heterogeneous causes only "make a difference" to a narrow subset of all cases of the disease and they are not causally or explanatorily relevant to most or all cases of the population-wide disease trait.

Notice how this problem does not arise if a disease is causally homogeneous. For diseases that have homogeneous causal factors, these factors can be targeted to explain all (or most) instances of the disease at the population level. This is because homogeneous causes do "make a difference" to all cases of the type-level effect. Homogeneous causes have causal and explanatory relevance of "broad scope." These causes can be targeted to explain a large percentage of all instances of the population-wide trait. In addition to this explanatory advantage, this feature is also present in the type of control that homogeneous causes have over type-level effects. Homogeneous causes can be targeted to control, prevent, and cure most or all instances of the disease in question. Again, this is because most of these instances have the same set of causes or causal etiology. Instead of aiming at a variety of heterogeneous causes, a single causal etiology can be targeted to achieve control over the population-wide disease. This helps reveal why it is valuable to identify diseases that meet this causal architecture and why there is a preference (and often assumption) that diseases have shared causal etiologies. Homogeneous causes provide a means of explaining and potentially controlling population-wide disease traits.

A second challenge posed by this type of causal complexity is that it introduces an additional question to be answered. This additional question is why do *different* causes all produce the *same* effect? Something about this situation seems puzzling and in need of further explanation. We find situations of causal heterogeneity puzzling because they conflict with a common intuition that similar effects should have similar causes.[18] When this assumption is not met, we expect some further explanation for why this is the case. This puzzle is similar to cases of "universality" in science, in which some "universal" behavior is produced or exhibited by systems with vastly different microstructural or causal details (Batterman 2002). For example, neurons with different physical details can exhibit the same firing behavior, and microstructurally distinct fluids can all exhibit similar features at their critical points (Ross 2015; Batterman 2002). We often find that these cases are puzzling and in need of further explanation. We want to know how the *same* behavior can be produced by systems with *different* microstructural details. This is similar to asking how the *same* type of disease can be

produced by *different* causes. Physicians and medical researchers often expect complete disease explanations to provide some satisfying answer to
these questions.

Consider an objection to these purported challenges. I have suggested
that heterogeneous causes fail to explain population-wide effects because
they have limited causal and explanatory relevance. If this is so, why not just
appeal to a disjunctive set of causal factors that together explain all (or
most) cases of the disease? One issue with this purported solution is that
there can be far too many causes to make this a feasible approach. Recall
that retinitis pigmentosa has anywhere from 75 to 300 causes and that some
psychiatric disorders are thought to have many more. Expecting scientists
to appeal to such a long list of factors is not a practical or realistic expectation, and it does not appear to reflect actual biological practice. We do not
find physicians and researchers explaining these conditions by citing
hundreds of distinct causal factors. Second, this makes gaining control over
the disease outcome much more difficult because of the vast number of
causes that a treatment or preventive strategy would need to target. Scientists explicitly mention this in the case of retinitis pigmentosa: "Given the
large number of mutations that cause RP, strategies of gene therapy aimed
at correcting each individual mutation may be an overwhelming task"
(Chang, Hao, and Wong 1993, 602). They claim that finding some shared
causal target "may be a much more practical approach because it would be
applicable to multiple mutations" and, thus, offer treatment for multiple
cases of the disease (Chang, Hao, and Wong 1993, 602). Third, this approach
still fails to address the extra question raised by this causal architecture—
namely, why do *different* causes all produce the *same* effect? Citing a disjunctive set of causes does not provide an answer to this question.

How do scientists address the challenges associated with causal heterogeneity? A first approach involves continuing to search for some shared
causal etiology that unifies the seemingly disparate heterogeneous causes.
One way of doing this involves identifying a "final common pathway" that
the upstream heterogeneous causes all converge on and operate through in
producing the disease of interest. In this case, the convergence point of the
final common pathway identifies some shared causal etiology for the disease. This shared etiology can be targeted to explain, control, and treat all (or
most) cases of the population-wide disease trait. For example, researchers
hypothesize that the process of apoptosis (or regulated cell death) may be
the final common pathway for the genetically heterogeneous disease retinitis

pigmentosa (Chang, Hao, and Wong 1993, 595). In light of this hypothesis, they suggest that apoptosis "is a logical target for intervention for a variety of retinal degenerations" (Chang, Hao, and Wong 1993, 601). Targeting this final common pathway would provide a way of treating many cases of retinitis pigmentosa, no matter what their most upstream genetic causes are. This approach has another advantage. It provides an answer to the question of why different causes produce the same effect. In this case, the explanation for this is that the many different causes all funnel through the same causal process, which ultimately leads to the singular effect of interest. This final common pathway identifies causes that do "make a difference" to all (or most) cases of the disease in question and causes that can be targeted to explain, predict, and control all or most cases of the disease at the population level. Alternatively, when a shared causal etiology cannot be found, a second approach is used. This second approach involves dividing up and redefining the disease trait on the basis of the heterogeneous causes. Thus, when researchers discovered that Parkinson's disease is caused by three different individually sufficient causes, some suggested that "there is no single Parkinson disease" and that this category represents "several different diseases" (Weiner 2008, 705; Stayte and Vissel 2014, 18). Both of these solutions restore causal homogeneity and the shared causal etiology standard. Furthermore, this second strategy reveals to what means medical researchers and physicians are willing to go to meet the causal homogeneity and shared causal etiology standards. They are willing to completely redefine disease traits.

5. DISEASE DISCOVERY AND CAUSATION: A FINAL COMPLICATING FEATURE

These four causal architectures provide categories and distinctions that can apply to scientific contexts more generally. They specify four different ways that causal factors can relate to an effect of interest. However, there are aspects of this chapter's analysis that pertain to medicine more exclusively. These aspects have to do with the fact that disease traits are often defined on the basis of their causes, while this is not always the case for other phenomena in science. This relates to an additional type of complexity involved in disease discovery and causation that deserves mention.

Recall the two-step process for disease discovery mentioned in section 2. The first step involves specifying some disease trait and its symptomology

(D), while the second step involves searching for the causal factors (C) that produce this trait. In this sense, "discovering" a disease involves uncovering both its symptomology *and* causal etiology. In addition to this discovery process, recall that the gold standard for defining disease traits involves defining them on the basis of their causes. Disease categories are expected to meet particular causal requirements. Diseases are defined on the basis of factors that (i) provide causal control over the disease and (ii) capture shared causal etiologies at the population level. If these conditions are not met, the legitimacy of the disease trait is questioned. Given this setup, consider the resulting dilemma. In order to search for the causes of a disease trait—and follow the established process of disease discovery—the trait in question first needs to be specified and defined. However, you cannot follow the gold standard way of defining the disease because this requires knowing what its causes are, and this is exactly what you are searching for. This captures a kind of catch-22 situation: you need to define diseases in order to search for their causes, but the best definitions of disease are supposed to reflect their causes. In other words, you need to define (D) to find (C), but the best definitions of (D) are supposed to reflect (C).

This situation forces psychiatrists and researchers to propose "best-guess" definitions of disease traits at the first step, before etiology is known at all. This captures how most of our current psychiatric disorders are conceptualized. The hope is that these "best guesses" will define diseases in ways that track causes that meet various causal requirements (i, ii) for disease. However, there is absolutely no guarantee that they will be able to do this. In fact, not only is there no guarantee of this, but the researchers' ability to ultimately find these causes is highly dependent on the first guess that they make. They might choose to define a disease by symptom clusters that have heterogeneous causes and that lack any shared causal etiology. If this is the case, it will be much harder to identify the shared causal process that produces this disease, if none exists. As Hyman states, if researchers "select study populations according to a system that is a poor mirror of nature, it is very hard to advance our understanding of psychiatric disease" (Casey et al. 2013, 810). In this sense, discovering these diseases is highly dependent on this first choice and on how diseases are initially defined. However, as captured in the preceding dilemma, this choice often needs to be made without knowing what the disease causes are.

Part of what this reveals is how disease classification influences causal discovery. Classification dictates where we shine the spotlight in searching for causes. If diseases are initially defined in ways that do not track shared causal etiologies, then this can complicate efforts to identify their causes. No matter how much you search among this group of patients, no shared causal process for their symptoms will be found because none exists. There are worries that the current classification system in psychiatry—the Diagnostic and Statistical Manual of Mental Disorders (DSM)—has defined diseases in ways that impede efforts to uncover their etiologies. In particular, there are worries that "these categories, based upon presenting signs and symptoms, may not capture fundamental underlying mechanisms of dysfunction" (Insel et al. 2010, 748). This leads researchers to refer to DSM disease categories as "diagnostic silos" and "epistemic blinders" that have not facilitated causal discovery yet continue to be used in searching for it (Hyman 2010; Casey et al. 2013, 811). Researchers worry that the continued use of these invalidated and "fictive" categories threatens to reify them as they continue to be used in diagnosis and experimental work (Casey et al. 2013, 811; Hyman 2010; Morris and Cuthbert 2012). Of course, we do not have proof that these categories lack shared causal etiologies—it may be that we just have not found them yet. However, the more we try to find these causes and the longer we go without making progress, the less likely this option seems.

Are there other ways to make progress in uncovering the etiologies of psychiatric disease? How should the field move forward? Some researchers caution against an approach of simply "replacing old flawed guesses with new guesses about disorder definitions" (Hyman 2010, 171). It is not clear that our next guess will be any better or that this approach is ideal for psychiatric disease. Another way forward involves inverting the disease discovery process. Instead of starting with an effect and searching for its causes, this solution involves starting with causes and searching for their effects. This strategy has been associated with the newer research domain criteria (RDoC) framework. This framework creates a "new kind of taxonomy for mental disorders" that focuses first on phenomena (or constructs) at different levels of analysis and only then on the disorders they link up with (Insel and Lieberman 2013). These levels of analysis include functional assessments of constructs at the level of genes, molecules, cells, neural circuits, physiology, behaviors, and self-reports (Morris and Cuthbert 2012, 31). The hope is that by starting with more concrete physical processes, helpful classifications might begin to emerge—classifications that may point to new disease divisions that are imperceptible within the current DSM framework. Instead of being constrained by DSM disease definitions, RDoC starts with potential functional impairments and tracks what downstream disease category they may lead to. As Casey and colleagues (2013) state, "a main way in which the RDoC project will influence neuroscience research is that rather than taking a diagnostic group and attempting to discover its underlying neurobiological basis, the RDoC approach uses our current understanding of behaviour–brain relationships as the starting point and relates these to clinical phenomenology." This allows researchers to start with potential upstream causes and search downstream for the effects, or disease categories, that they lead to.

While RDoC involves an inversion of the traditional disease discovery process, it is likely that progress in unveiling the etiologies of psychiatric disease will involve more of a back-and-forth process. Researchers might start with causal factors, search for their effects, and change how they isolate causes on the basis of what they find. Alternatively, they might start with a new disease category, search for its causes, and redefine the category on the basis of what they uncover. Where the goal is to link up particular causal processes with particular disease definitions, researchers will likely toggle back and forth between both causes and effects until they find the right match.

6. CONCLUSION

This chapter has clarified four causal architectures and two types of causal complexity that are common in psychiatric genetics. Multicausality and causal heterogeneity capture distinct types of complexity at the level of disease causation. This chapter has examined how these types of causal complexity should be understood, how they challenge disease explanation, and how scientists are working to overcome these challenges. While aspects of this analysis pertain to biomedicine more specifically, these four causal architectures are likely to provide general categories and distinctions that apply to scientific contexts more broadly.

This analysis provides a different way to understand common claims that psychiatry is in a "crisis." First, it helps clarify why explanation in psychiatry is difficult without suggesting that psychiatric diseases are "intractable," "enigmatic," or "beyond scientific understanding." In fact, this analysis reveals how scientific progress is being made in this domain. It reveals the sound methodologies that guide efforts to discover, understand, and explain psychiatric diseases, and it indicates that these methodologies are found in other medical subfields. Relatedly, it shows that these types of complexity are not unique to psychiatry but that they are found in other areas of medicine. This is seen in disease examples such as PKU and retinitis pigmentosa, which exhibit multicausality and causal heterogeneity, respectively, despite being viewed as physical medicine or nonpsychiatric diseases. Instead of being viewed as a defunct discipline in "crisis," psychiatry is better understood as a field at the forefront of disease discovery and that seeks to uncover and shed light on some of the more challenging diseases that confront all of medicine.[19] In many ways, psychiatry is a "safe haven" for diseases of unknown etiology—it allows these diseases to be taken seriously, scrutinized, and modified so that they can be fairly judged on the basis of various foundational standards in modern medicine. The constant push for etiological understanding in psychiatry leads researchers in this area to explicitly reflect on the abstract principles that diseases are expected to meet—why such principles are important and when (if ever) they should be relaxed. These reflections reveal standards and methodologies that are not just found in psychiatry but that are present throughout medicine more generally.

NOTES

1. I follow the custom of referring to both signs and symptoms as "symptomology."

2. For further discussion of this disease discovery process, see Ross 2018.

3. As Insel et al. (2010) state, "History shows that predictable problems arise with early, descriptive diagnostic systems designed without an accurate understanding of pathophysiology. Throughout medicine, disorders once considered unitary based on clinical presentation have been shown to be heterogeneous. . . . Conversely, history also shows that syndromes appearing clinically distinct may result from the same etiology" (748).

4. This mirrors a strategy that originated with classical genetics, which involves starting from a phenotype and searching for its genetic causes ("forward genetics"), as opposed to starting from gene variants and searching for their effects ("reverse genetics") (Lawson and Wolfe 2011).

5. In other words, they have to do with simplicity and complexity at the level of causes (given some effect) and not at the level of an effect (given some cause).

6. In other words, knowing that a *single instance* of some disease is multicausal or monocausal provides no information about whether *all instances* of the disease are causally heterogeneous or homogeneous. A disease that is monocausal and causally homogeneous is a disease that has a single main cause, where this cause produces all cases of the population-wide disease. These diseases fit the standard "hard" medical model and they include examples such as scurvy, tuberculosis, Huntington's disease, and chicken pox, to name a few. A disease that is monocausal and causally heterogeneous is one in which a single main cause produces each instance of the disease (monocausality) but different single causes produce distinct cases of the same disease (causal heterogeneity). A disease can be multicausal and causally homogeneous if each instance of the disease is caused by multiple factors (multicausality), but the same combination of factors causes every instance of the disease (causal homogeneity). Phenylketonuria (PKU) is an example of this because two factors cause each instance of the disease (multicausality) and the same two factors cause all cases of the disease (causal homogeneity). Finally, a disease can be multicausal and causally heterogeneous if each instance of the disease is caused by many factors (multicausality) but there are different combinations of causal factors that produce distinct instances of the disease (causal heterogeneity). These points are discussed in more detail throughout the chapter.

7. These include manipulations of the tubercle bacteria, chicken pox virus, *Giardia* parasite, and *huntingtin* gene variant, respectively.

8. They actually have many special types of control over disease (Ross forthcoming), but I focus on one type that helps clarify what is meant by multicausality.

9. This notion of probable causal control is similar to Cheng's (1997) notion of "causal power."

10. The fact that this probable control holds across a wide range of background conditions is also related to Woodward's (2010) notion of stability and Kendler's (2005) notion of noncontingency of association.

11. Discussions of this type of causal complexity are found in the philosophical literature. This is seen in Mitchell's (2008, 24) discussion of situations where there are "multiple causes additively or interactively contributing to the production of a major effect."

12. For more on this, see Ross forthcoming.

13. In other words, "etiologic heterogeneity refers to a phenomenon that occurs in the general population when multiple groups of disease cases,

such as breast cancer clusters, exhibit similar clinical features, but are in fact the result of differing events or exposures" (Hernandez and Blazer 2006, 46).

14. These include high blood pressure, hyperlipidemia, retinitis pigmentosa, Alzheimer's disease, cystic fibrosis, lipoprotein lipase, and polycystic kidney disease.

15. There is a further distinction between locus and allelic heterogeneity. Locus heterogeneity refers to mutations at different loci (or in different genes) that are capable of producing the same outcome. Allelic heterogeneity indicates that different mutations (or alleles) at the same gene produce the same outcome. My analysis focuses on locus heterogeneity.

16. As Stegenga (2018, 67) states, targeting these causes can "at best improve the health of a subset of people" with the disease in question.

17. However, notice that the control is uneven—C_1 can be used to reliably cause disease, but not to reliably prevent it. The causal framework that I rely on requires that causes have control over both contrasts of the explanatory target—namely, the presence and absence of the disease.

18. We see this assumption in Hume 1738, for example.

19. Ironically, although psychiatry is sometimes criticized for only housing diseases of unknown etiology, once the etiology of some "psychiatric" conditions is uncovered, the resulting disease is often relocated to another area of medicine, such as neurology. This has occurred with various forms of dementia, which are now considered "neurologic" as opposed to "psychiatric" in nature. This can prevent psychiatry from receiving full credit for disease discovery. It seems unreasonable to criticize psychiatry for only (or mainly) dealing with diseases of unknown etiology while at the same time recognizing that diseases are removed from this field once their etiologies are identified.

REFERENCES

Barondes, S. H. 1992. "How Genetically Heterogeneous Are the Major Psychiatric Disorders?" *Journal of Psychiatric Research* 26, no. 4: 1–5.

Batterman, R. W. 2002. *The Devil in the Details: Asymptotic Reasoning in Explanation, Reduction, and Emergence.* Oxford: Oxford University Press.

Betancur, C. 2011. "Etiological Heterogeneity in Autism Spectrum Disorders: More Than 100 Genetic and Genomic Disorders and Still Counting." *Brain Research* 1380, no. C: 42–77.

Carr, S. M. 2014. *Penetrance versus Expressivity*. https://www.mun.ca/biology /scarr/Penetrance_vs_Expressivity.html.

Casey, B. J., N. Craddock, B. N. Cuthbert, S. E. Hyman, F. S. Lee, and K. J. Ressler. 2013. "DSM-5 and RDoC: Progress in Psychiatry Research?" *Nature Reviews Neuroscience* 14, no. 11: 810–14.

Caspi, A., and T. E. Moffitt. 2006. "Gene–Environment Interactions in Psychiatry: Joining Forces with Neuroscience." *Nature Reviews Neuroscience* 7, no. 7: 583–90.

Chang, G.-Q., Y. Hao, and F. Wong. 1993. "Apoptosis: Final Common Pathway of Photoreceptor Death in rd, rds, and Rhodopsin Mutant Mice." *Neuron* 11, no. 4: 595–605.

Cheng, P. W. 1997. "From Covariation to Causation: A Causal Power Theory." *Psychological Review* 104, no. 2: 367–405.

Cooper, D. N., M. Krawczak, C. Polychronakos, C. Tyler-Smith, and H. Kehrer-Sawatzki. 2013. "Where Genotype Is Not Predictive of Phenotype: Towards an Understanding of the Molecular Basis of Reduced Penetrance in Human Inherited Disease." *Human Genetics* 132: 1077–130.

Egger, G. 2012. "In Search of a Germ Theory Equivalent for Chronic Disease." *Preventing Chronic Disease* 9, no. E95.

Engel, G. L. 1977. "The Need for a New Medical Model: A Challenge for Biomedicine." *Science* 196, no. 4286: 129–36.

Goldstein, D. B. 2009. "Common Genetic Variation and Human Traits." *New England Journal of Medicine* 360, no. 17: 1696–98.

Griffiths, A., S. R. Wessler, R. Lewontin, and S. B. Carroll. 2008. *Introduction to Genetic Analysis, Volume 9*. New York: W. H. Freeman and Company.

Hernandez, L. M., and D. G. Blazer. 2006. *Genes, Behavior, and the Social Environment: Moving beyond the Nature/Nurture Debate*. Washington, DC: The National Academies Press.

Hoffecker, J. F. 2011. *Landscape of the Mind: Human Evolution and the Archaeology of Thought*. New York: Columbia University Press.

Hume, D. 1738. *A Treatise of Human Nature*. Book 1. New York: Oxford University Press Warehouse.

Hyman, S. E. 2010. "The Diagnosis of Mental Disorders: The Problem of Reification." *Annual Review of Clinical Psychology* 6, no. 1: 155–79.

Hyman, S. E. 2013. "Psychiatric Drug Development: Diagnosing a Crisis." *Cerebrum* 2013: 5.

Ideker, T., J. Dutkowski, and L. Hood. 2011. "Boosting Signal-to-Noise in Complex Biology: Prior Knowledge Is Power." *Cell* 144, no. 6: 860–63.

Insel, T., B. Cuthbert, M. Garvey, R. Heinssen, D. Pine, K. Quinn, C. Sanislow, and P. Wang, P. 2010. "Research Domain Criteria (RDoC): Toward a New Classification Framework for Research on Mental Disorders." American Journal of Psychiatry 167, no. 7: 748–51.

Insel, T. R., & Lieberman, J. A. 2013. "DSM-5 and RDoC: Shared Interests." The National Institute of Mental Health, http://www.nimh.nih.gov/news/science-news/2013/dsm-5-and-rdoc-shared-interests.shtml.

Kendler, K. S. 2005. "'A Gene for . . .': The Nature of Gene Action in Psychiatric Disorders." American Journal of Psychiatry 162: 1243–52.

Kendler, K. S. 2012. "Levels of Explanation in Psychiatric and Substance Use Disorders: Implications for the Development of an Etiologically Based Nosology." Molecular Psychiatry 17, no. 1: 11–21.

Kendler, K. S. 2013. "What Psychiatric Genetics Has Taught Us about the Nature of Psychiatric Illness and What Is Left to Learn." Molecular Psychiatry 18, no. 10: 1058–66.

Kendler, K. S., and P. Zachar. 2008. "The Incredible Insecurity of Psychiatric Nosology." In Philosophical Issues in Psychiatry: Explanation, Phenomenology, and Nosology, 368–82. Baltimore: Johns Hopkins University Press.

Lawson, N. D., and S. A. Wolfe. 2011. "Forward and Reverse Genetic Approaches for the Analysis of Vertebrate Development in the Zebrafish." Developmental Cell 21, no. 1: 48–64.

Lemoine, M. 2016. "Molecular Complexity." In Philosophy of Molecular Medicine, edited by G. Boniolo and M. J. Nathan, 81–99. New York: Routledge.

Maier, R. M., P. M. Visscher, M. R. Robinson, and N. R. Wray. 2017. "Embracing Polygenicity: A Review of Methods and Tools for Psychiatric Genetics Research." Psychological Medicine 48: 1055–67.

McGinniss, M. J., and M. M. Kaback. 2013. "Heterozygote Testing and Carrier Screening." In Emery and Rimoin's Principles and Practice of Medical Genetics, 1–10. Amsterdam: Elsevier.

Mitchell, K. J. 2012. "What Is Complex about Complex Disorders?" Genome Biology 13, no. 237: 1–12.

Mitchell, S. D. 2008. "Explaining Complex Behavior." In Philosophical Issues in Psychiatry, 21–36. Baltimore: Johns Hopkins University Press.

Morgan, A. 2015. "Is Psychiatry Dying? Crisis and Critique in Contemporary Psychiatry." Social Theory & Health 13, no. 2: 141–61.

Morris, S. E., and B. N. Cuthbert. 2012. "Research Domain Criteria: Cognitive Systems, Neural Circuits, and Dimensions of Behavior." Dialogues in Clinical Neuroscience 14, no. 1: 29–37.

Murphy, D. 2006. *Psychiatry in the Scientific Image.* Cambridge, Mass.: MIT Press.

Murphy, E. A. 1997. *The Logic of Medicine,* 2nd ed. Baltimore: Johns Hopkins University Press.

Nakagawa, S., and I. C. Cuthill. 2007. "Effect Size, Confidence Interval and Statistical Significance: A Practical Guide for Biologists." *Biological Reviews* 82, no. 4: 591–605.

Nandipati, S., and I. Litvan. 2016. "Environmental Exposures and Parkinson's Disease." *International Journal of Environmental Research and Public Health* 13, no. 9.

Park, J.-H., S. Wacholder, M. H. Gail, U. Peters, K. B. Jacobs, S. J. Chanock, and N. Chatterjee. 2010. "Estimation of Effect Size Distribution from Genome-Wide Association Studies and Implications for Future Discoveries." *Nature Genetics* 2: 570–75.

Plomin, R., and Y. Kovas. 2005. "Generalist Genes and Learning Disabilities." *Psychological Bulletin* 131, no. 4: 592–617.

Poland, J., and S. Tekin. 2017. "Introduction: Psychiatric Research and Extraordinary Science." In *Extraordinary Science and Psychiatry.* Cambridge, Mass.: MIT Press.

Price, A. L., C. C. A. Spencer, and P. Donnelly. 2015. "Progress and Promise in Understanding the Genetic Basis of Common Diseases." *Proceedings of the Royal Society* 282, no. 1821.

Ross, L. N. 2015. "Dynamical Models and Explanation in Neuroscience." *Philosophy of Science* 82, no. 1: 32–54.

Ross, L. N. 2018. "The Doctrine of Specific Etiology." *Biology & Philosophy* 33, no. 37.

Ross, L. N. Forthcoming. "Causal Control: A Rationale for Causal Selection." In *Philosophical Perspectives on Causal Reasoning in Biology,* edited by B. Hanley, C. K. Waters, and J. Woodward. Minneapolis: Minnesota Studies in Philosophy of Science.

Stayte, S., and B. Vissel. 2014. "Advances in Non-Dopaminergic Treatments for Parkinson's Disease." *Frontiers in Neuroscience: Neuropharmacology* 8: 1–29.

Stegenga, J. 2018. *Medical Nihilism.* Oxford: Oxford University Press.

Stewart, A., P. Brice, H. Burton, P. Pharoah, S. Sanderson, and R. Zimmern. 2007. *Genetics, Health Care and Public Policy.* Cambridge: Cambridge University Press.

Sullivan, P. F., M. J. Daly, and M. O'Donovan. 2012. "Genetic Architectures of Psychiatric Disorders: The Emerging Picture and Its Implications." *Nature Reviews Genetics* 13, no. 8: 537–51.

Takahashi, S. 2013. "Heterogeneity of Schizophrenia: Genetic and Symptomatic Factors." *American Journal of Medical Genetics* 162, no. 7: 648–52.

Torkamani, A., N. E. Wineinger, and E. J. Topol. 2018. "The Personal and Clinical Utility of Polygenic Risk Scores." *Nature Reviews Genetics* 19: 581–90.

Tsuang, M. T., S. J. Glatt, and S. V. Faraone. 2006. "The Complex Genetics of Psychiatric Disorders." In *Principles of Molecular Medicine,* 1184–90. Totowa, NJ: Humana Press.

Uher, R., and A. Zwicker. 2017. "Etiology in Psychiatry: Embracing the Reality of Polygene-Environmental Causation of Mental Illness." *World Psychiatry* 16, no. 2: 121–29.

Wang, K., M. Li, and H. Hakonarson. 2010. "Analysing Biological Pathways in Genome-Wide Association Studies." *Nature Reviews Genetics* 11: 843–54.

Weiner, W. J. 2008. "There Is No Parkinson Disease." *Archives of Neurology* 65, no. 6: 705–8.

Weiss, K. M. 2007. "Phenotype and Genotype." In *Keywords and Concepts in Evolutionary Developmental Biology,* 297–87. New Delhi: Discovery Publishing House.

Woodward, J. 2003. *Making Things Happen.* Oxford: Oxford University Press.

Woodward, J. 2010. "Causation in Biology: Stability, Specificity, and the Choice of Levels of Explanation." *Biology & Philosophy* 25: 287–318.

Zachar, P. 2014. "Beyond Natural Kinds: Toward a 'Relevant' 'Scientific' Taxonomy in Psychiatry." In *Classifying Psychopathology: Mental Kinds and Natural Kinds,* edited by H. Kincaid and J. Sullivan, 75–104. Cambridge, Mass.: MIT Press.

9

THE GROUNDED FUNCTIONALITY
ACCOUNT OF NATURAL KINDS

MARC ERESHEFSKY AND THOMAS A. C. REYDON

1. INTRODUCTION

What are natural kinds? In addressing this question, philosophers have started from various interpretations of what the problem of natural kinds is—a metaphysical question about the fundamental building blocks of the world, a question about the reference of substance terms in everyday language, an epistemological question about the basis of inductive inferences, and so on. Answers to these questions radically diverge. Some philosophers posit robustly metaphysical accounts of kinds, positing that natural kinds have essences (Ellis 2001) or that they are universals (Hawley and Bird 2011). Other accounts of natural kinds emphasize the causal nature of such kinds—they either are based upon causal mechanisms (Boyd 1999b) or are nodes in causal networks (Khalidi 2018). Still others offer less metaphysical accounts of natural kinds. Such accounts see kinds as clusters of co-occurring properties (Slater 2015) or as groupings that we are forced to use in support of inductive and explanatory practices (Magnus 2012) or as those groupings identified by converging epistemic practices (Franklin-Hall 2015).

In this chapter, we argue that philosophical theories of natural kinds are insufficiently focused on classificatory practice in science. Available theories of natural kinds tend to suffer from two defects. First, some of those theories are developed according to a priori considerations (Reydon 2010a; 2010b; 2014; Ereshefsky 2018). As we will show, the result is that such theories of natural kinds fail to help us understand why classificatory practices in science are successful. Furthermore, theories of natural kinds are frequently overarching theories—they claim that all natural kind classifications are posited to capture the same universal aim, such as highlighting the causal structure of the world. However, we argue that scientists have a variety of reasons for positing natural kind classifications and extant philosophical theories fail to capture that variety. Given these features of philosophical accounts of natural kinds—their a priori basis and/or their overarching nature—there is a discrepancy between the philosophical literature on natural kinds and classificatory practices in science.

This chapter highlights the preceding problems, and it offers an account of natural kinds that better reflects classificatory practices in science. We call this account the "Grounded Functionality Account of natural kinds," or GFA for short. On the one hand, this account is attentive to the local practices of classificatory projects. On the other hand, it offers two constraints on natural kind classifications, namely that such classifications serve the epistemic (as well as nonepistemic—see Reydon and Ereshefsky 2022) functions they are posited for and that they satisfy those functions because they are grounded in the world. The GFA, in other words, suggests that natural kind classifications help scientists achieve various aims and that success is due to those classifications properly capturing some aspects of the world. Which aspects of the world natural kinds classifications should be grounded in will vary dramatically between different research contexts, and that diversity makes sense given the variety of aims scientists have for constructing classifications.

As mentioned earlier, there are various accounts of natural kinds in the philosophical literature. Those accounts tend to be positive accounts of natural kinds—they give an account of the nature of natural kinds. But there are also negative or skeptical accounts. Hacking (2007b) and Ludwig (2018), for instance, argue that philosophical research on natural kinds has been fruitless and philosophers should stop trying to develop theories of natural kinds. We believe that the GFA blunts such skepticism. The GFA offers an account of natural kinds that is attentive to the variety of reasons scientists have for positing natural kind classifications. At the same time, it highlights a positive philosophical project that philosophers and scientists engage in when they think about natural kinds and classification.

What follows is broken into four parts. The next section of this chapter illustrates that philosophical accounts of natural kinds tend to be overly detached from actual classificatory practice in science. In that section, we answer the question of why a new account of natural kinds is needed. The third section suggests a way to better align a theory of natural kinds with the diverse epistemic aims of scientists and presents the main aspects of our account of natural kinds, the GFA. In the section that follows, we flesh out some of the details of our account and tackle the vexing question: What makes a natural kind natural? In the final section, we turn to recent skepticism concerning natural kinds and suggest that our account blunts that skepticism.

2. PHILOSOPHICAL ACCOUNTS OF NATURAL KINDS ARE NOT NATURALISTIC ENOUGH

As mentioned in the introduction, we are concerned that philosophical accounts of natural kinds are too divorced from actual classificatory practice to be relevant to that practice. In our investigation of natural kind theories we start from the assumption that philosophical theories of natural kinds should be relevant to successful classificatory practices in science. There are various ways in which a philosophical theory of natural kinds can be relevant to classificatory practices in science. Consider two straightforward— and, we think, uncontroversial—desiderata of natural kind theories. One is that we would like a philosophical approach to natural kinds to help us understand why classificatory practices in science are successful. That is, we'd like a philosophical analysis of natural kinds to tell us why certain classifications help achieve the epistemic and nonepistemic aims of the scientists that use them. Another desideratum of a philosophical theory of natural kinds is that we would like such a theory to give us some guidance in determining whether a classification is indeed a classification of natural kinds and to distinguish natural kinds from other kinds. An account of natural kinds should have some normative force and give some guidance in telling us whether a classification is a good candidate or a poor candidate for being a natural kind classification.

In the philosophical literature, there is a class of philosophical accounts of natural kinds that does not meet either of these desiderata. Here we have in mind philosophical theories of natural kinds that are developed on primarily a priori grounds. Such accounts of natural kinds are not developed by observing and learning from actual classificatory practices in science but are developed on the basis of a priori considerations and intuitions. Such accounts often come from the philosophical school of analytic metaphysics. One problem with such approaches is their reliance on intuitions. A well-known example of such analytic metaphysics is Putnam's (1975) argument for natural kind essentialism involving his infamous Twin Earth thought experiment. Many philosophers have noted that intuitions can be misleading when it comes to metaphysics (see, for instance, Callender 2011; Papineau 2015; Bryant 2017). That is certainly a major concern. But our target here is that such a priori approaches to natural kinds do not help us understand the success of natural kind classifications in science, nor do they provide guidance in judging whether a classification is a classification of natural kinds.

Consider the debate among philosophers who believe that natural kinds are universals. Such philosophers disagree over the appropriate type of universals that natural kinds are thought to be. Lowe (2006), for example, maintains that natural kinds are substantial universals, as does Ellis (2001). For Lowe, substantial universals are an irreducible type of ontological category in his four-category ontology. He believes that natural kinds are a fundamental part of our universe, whereas properties are nonsubstantial universals. According to Lowe, natural kinds are substantial universals characterized by properties. For instance, the kind *water* is a substantial universal characterized by the property of being H_2O. Hawley and Bird (2011) also hold that natural kinds are universals. However, they think that natural kinds are complex universals rather than substantial universals. Complex universals, they suggest, are universals whose parts are universals. They offer the example of the kind *electron,* which is a complex universal consisting of the universals of an electron's mass, charge, and spin.

We won't go into any further details of the debate among philosophers that hold that natural kinds are universals. Instead, we want to highlight that their debate about the nature of natural kinds is so abstract that their a priori theories of natural kinds neither illuminate successful classificatory practices nor give guidance in how to conduct such practices. Take, for example, the shift in taxonomic practices concerning biological species in the early twentieth century that shifted from a morphological approach to focusing on interbreeding. The morphological approach uses morphological similarity to sort organisms into species, while the interbreeding approach does that on the basis of which organisms can successfully interbreed and produce fertile offspring. The interbreeding approach allowed biologists to more accurately sort organisms into species than the morphological approach. The morphological approach incorrectly sorts similar males in different species into the same species, whereas the interbreeding approach revealed that those males belong to different species (Ridley 1993). When considering this case of taxonomic progress, one might ask if the debate over whether natural kinds are substantial universals or complex universals has any relevance to it. Knowing that natural kinds are a particular type of universal does not help us understand why the interbreeding approach is more successful than the morphological approach. Furthermore, knowing that natural kinds are a particular type of universal does not help us judge whether the interbreeding approach is a better approach to biological species than the morphological approach. There is, we submit, a significant disconnect between a priori theories of natural kinds and classificatory practice in science. As indicated by this case of species, a priori approaches to natural kinds do not illuminate why certain classificatory practices in science have been progressive. Similarly, such a priori approaches to natural kinds provide no guidance in discriminating between natural kind classifications and nonnatural kind classifications in actual taxonomic practice.

Let's turn to another and more recent a priori account of natural kinds: Franklin-Hall's (2015) "Categorical Bottleneck" account of natural kinds. Franklin-Hall locates natural kinds at the intersection of investigations conducted by different epistemic agents. In particular, she writes that "natural kinds are groupings that match those categories that well serve actual inquirers along with (what I call) 'neighboring agents'—those different somewhat from actual inquirers in their particular epistemic aims and cognitive capacities" (2015, 940). A virtue of Franklin-Hall's account is that it highlights the role intersubjectivity plays in identifying natural kinds. However, the sort of intersubjectivity that Franklin-Hall requires is too a priori and too distant from actual classificatory practice in science. The sort of intersubjectivity her account employs turns on "neighboring agents" where those neighboring agents are, as seen in the preceding quote, not "actual inquirers." Such "neighboring agents" are *possible* inquirers who occupy positions in what Franklin-Hall (2015, 940) calls an "epistemic agent space"—that is, a conceptual space with all possible epistemic aims and cognitive capacities as its dimensions and in which all possible inquirers occupy specific locations. Natural kinds, in Franklin-Hall's account, then are identified as those kinds that robustly continue to serve the aims of inquirers under comparatively small movements in "epistemic agent space" toward slightly different aims or cognitive capabilities. But by relying on an abstract "epistemic agent space" and possible inquirers, Franklin-Hall's account is not an account of what natural kind classifications are in actual scientific practice but an a priori, otherworldly account of natural kinds. Indeed, Franklin-Hall's account is nonoperational: How could we check that nonactual inquirers would pick out the same kinds as actual inquirers? Just as in the case of universalist natural kind theories, Franklin-Hall's account is too distant from actual classificatory practice to illuminate such practices: relying on nonactual epistemic agents does not help us understand the success of actual classificatory practices. Moreover, relying on nonactual epistemic agents fails to give guidance in choosing among real classificatory practices.

We have seen that a priori approaches to natural kinds tend to be too distant and irrelevant to actual classificatory practice in science to be useful for understanding how science works. This is a big strike against them. There are of course other approaches to natural kinds that attempt to be more naturalistic and rely less on a priori and intuitive reasoning. However, many of these approaches also fail to capture actual classificatory work in science, and they do so for a different reason. A standard feature of many accounts of natural kinds is that they are overarching accounts of natural kinds. That is, they are theories of the form "All natural kinds have some feature X." Proponents of such accounts disagree on what "X" refers to, but they tend to agree that philosophical accounts of natural kinds should be overarching accounts that apply to all natural kind classifications throughout the sciences. We don't take issue with theories of natural kinds being overarching theories. Our concern is that such overarching theories of natural kinds neglect large swaths of classificatory practice in science. If a philosophical account of natural kinds neglects large parts of classification in science, then it is of little help in understanding many parts of classificatory practice in science. Let's consider some prevalent philosophical accounts of natural kinds that do just that—they neglect large parts of classificatory practice in science.

One overarching criterion often placed on natural kinds by philosophers is that such kinds should be causal kinds (e.g., Boyd 1999a; 1999b; 2003; Wilson, Barker, and Brigandt 2007; Samuels 2009; Craver 2009; Khalidi 2013; 2018). According to this criterion, the members of a natural kind should share a similar set of causal components, mechanisms, or nodes. Boyd, for instance, talks of "causal structures" (1999a, 159) and "homeostatic mechanisms" (1999a, 165), while Khalidi talks in terms of "clusters of causal properties" (2018). Despite the enthusiasm among philosophers for capturing the causal structure of the world, a significant number of scientists produce classifications that do not aim to capture causal kinds (see Ereshefsky and Reydon 2015).

Microbiologists, for example, construct classifications of microbial kinds, but not with the aim of capturing the causal structure of the world. Instead they aim to posit classifications of microbial kinds that are stable

and readily identifiable. Why? Because identifying kinds with such properties is vital for research in microbiology and medicine. If, for instance, a bac-
teriologist is studying the relations among bacteria within a biofilm, she
needs to refer to a stable and readily identifiable set of microbial kinds. The
same applies to the medical researcher that studies bacteria in our digestive
system. The most widely accepted approach to bacterial kinds, the Phylo-
Phenetic Species Concept (PPSC) (Rosselló-Mora and Amann 2001; Stackebrandt 2006), uses several types of genetic markers to identify bacterial
kinds. Those markers are chosen not because they capture the causal structure of the world or any causal mechanisms in microbes but because they
provide stable and readily identifiable groups of microbes. As Stackebrandt
(2006, 36–37) writes, "bacteriologists in particular follow guidelines and
recommendations that provide stability, reproducibility, and coherence in
taxonomy." Though many philosophers are keen on science revealing the
causal structure of the world, the pursuit of causal kinds is not of interest to
these microbiologists. Those biologists use genetic markers to identify and
reidentify groups of organisms in the world such that taxonomy in bacteri-
ology is both doable and stable. While some may suggest that causes always
lie in the background of epistemic and nonepistemic aims, our point is that
the PPSC does not refer to causes: causes do not play a role when sorting
microbes into species using the PPSC. Philosophers that maintain that natural kinds are causal kinds offer an approach to natural kinds that is irrelevant to the taxonomic work of these biologists. In other words, there is a
mismatch between the philosophical desideratum that natural kinds be
causal and the reasons many biologists have for positing natural kind
classifications.

Let's turn to another requirement that is commonly placed on natural
kinds by philosophers, namely, that the members of a natural kind should
share numerous co-occurring properties such that natural kind classifications can underwrite induction. Many philosophers hold this assumption
(Boyd 1999a; 1999b; 2003; Lowe 2006; Wilson, Barker, and Brigandt 2007;
Hawley and Bird 2011; Magnus 2012; Khalidi 2013; 2018; Slater 2013; 2015),
which goes back to the British Empiricists, especially Mill's *System of Logic*.
But despite the popularity of this assumption, many scientific classifications do not highlight inductive kinds (see Ereshefsky and Reydon 2015).

Consider the kinds of biological taxonomy. One aim of biological taxonomy is to identify branches on the tree of life. Taxa, such as species and
genera, are considered branches on the tree of life. Such taxonomic kinds are

first and foremost historical entities and only secondarily groups of organisms with numerous similarities (Ereshefsky 2001). The challenge for those that assert that natural kinds are groups of entities with numerous similarities is that classifying by similarity and classifying by history can conflict. And when they do conflict, the view that natural kinds are inductive kinds fails to capture the classificatory practices of those biologists that classify by history.

As an example, branching on the tree of life frequently occurs through allopatric speciation—when one population becomes geographically separated from the rest of a species and gradually evolves into a new species. When a population branches off from its ancestor species, the organisms of both the isolated population and the ancestral branch continue for a while to have the same family of properties. Splitting need not be accompanied by immediate changes in traits, and often traits remain conserved over considerable evolutionary time scales, such that two different branches on the tree of life contain organisms that are overwhelmingly similar (Reydon 2006). If we follow the philosophical position that kinds are inductive kinds, we should consider the new branch and the ancestral branch as constituting one species, given that their organisms share a large number of properties. Yet generally recognized models of speciation hold that when an isolated population branches off from its ancestral species, speciation occurs (Coyne and Orr 2004). In short, the aim of biological taxonomy is to classify distinct branches on the tree of life rather than clusters of similar organisms. Biologists interested in classifying the tree of life reject the common philosophical assumption that all natural kinds should be inductive kinds.

From this example, we see that the philosophical assumption that natural kinds are inductive kinds is inconsistent with some classificatory practices in biology. From the earlier example concerning microbiology, we see that the philosophical assumption that natural kinds are causal kinds is also inconsistent with some classificatory practices in biology. Putting these together we see a pattern. Philosophers promote all-encompassing accounts of natural kinds: *all* natural kinds in science should be causal, or *all* natural kinds in science should be inductive. However, such overarching accounts of natural kinds are inconsistent with highly successful classificatory practices in science. Note that we are not denying that some classifications in science underwrite inductions, and some are causal kinds. We are merely pointing out that the tendency of philosophers to propose overarching accounts of natural kinds is mistaken: universal approaches to successful classifications in science fail to capture the breadth of classificatory practices in science.

There are other overarching requirements that philosophers place on natural kinds besides the requirements that natural kinds be causal kinds or inductive kinds. Consider some of the criteria listed by Bird and Tobin (2017): that all natural kinds are mind-independent, that natural kinds should form hierarchies, and that natural kinds should be categorically distinct. Each of these requirements is inconsistent with some successful classificatory practice in science. As we will see later, the requirement that natural kinds be mind-independent is inconsistent with classifications in the human, social, and medical sciences. The requirement that natural kinds be hierarchically arranged conflicts with classificatory practices in chemistry (Hendry 2010; Khalidi 2013). And the requirement that natural kinds be categorically distinct, that is, not bleed into one another, is violated in some areas of biology (Ereshefsky 2001).

The problem with many philosophical theories of natural kinds is not merely that those theories have counterexamples. It is more pressing than that. If philosophical research on natural kinds is supposed to provide an understanding of our classificatory practices, then such research should learn from our best classificatory practices. By failing to capture the array of epistemic reasons scientists have for positing natural kind classifications, available theories of natural kinds fail to provide an understanding of many classificatory practices in science. Couple that problem with the one we saw earlier, namely that many philosophical approaches to natural kinds are a priori and too removed from actual classificatory practice, and we see that a more practice-oriented account of natural kinds is needed. In what follows, we suggest such an account.

3. BALANCING NATURALISM AND NORMATIVITY

The account of natural kinds we offer is in part inspired by Laudan's (1987; 1990) normative naturalism and Woodward's (2014) functional account of causal reasoning. Laudan developed normative naturalism for evaluating the methodological rules of a research tradition. According to Laudan, science consists of research traditions, which contain theories, methodological rules, and overall aims. By "methodological rule," Laudan means such rules as prefer simpler theories or prefer more unified theories over less unified ones. According to normative naturalism, scientists should adopt those methodological rules that best promote the aims of their research tradition. His normative naturalism is naturalistic in that the actual aims of a discipline (rather than some philosophical abstraction of science or an ideal of what science should be) are used to judge which methodological rules to use. It is normative in a goal-directed sense because there are norms for evaluating methodological rules. Woodward's functional account of causal reasoning works in a similar fashion. According to Woodward, different types of causal reasoning are used to achieve different epistemic goals. He suggests that a type of causal reasoning should be judged by how well it helps achieve the epistemic goal it was posited for. As he writes, "causal information and reasoning are sometimes useful or functional in the sense of serving various goals and purposes that we have," such that talking about causes is best seen as "a kind of epistemic technology—as a tool—and, like other technologies, judged in terms of how well it serves our goals and purposes" (Woodward 2014, 693–94). Woodward's account is naturalistic because scientists' actual epistemic goals (rather than metaphysical views about what causes are) are used to judge types of causal reasoning. It is normative in a goal-directed way because a type of causal reasoning is evaluated by how well it satisfies the particular epistemic aim it was posited for.

We would like to suggest an approach to natural kinds that is similar in spirit, which we call the Grounded Functionality Account of natural kinds (GFA). It is a functional approach because on the GFA a natural kind classification is judged by how well it functions in achieving the epistemic aims (or nonepistemic aims—see later) it is posited for.[1] Call this the "functionality condition" on natural kinds. (We will discuss the "grounded" aspect of the GFA in the next section.) We can illustrate how the GFA works by using the notion of a *classificatory program* (Ereshefsky 2001; Ereshefsky and Reydon 2015). Classificatory programs are analogous to Laudan's research traditions and consist of three parts: classifications, motivating principles, and sorting principles. The classifications produced by a classificatory program highlight putative natural kinds. Sorting principles sort entities into kinds. Motivating principles are the aims of a classificatory program and motivate why that program should sort entities a particular way. In science, natural kind classifications are posited for an array of epistemic as well as nonepistemic reasons. According to the GFA, a natural kind classification should be evaluated by how well it satisfies the aims of *its specific* classificatory program.

As an example of a classificatory program, consider Mayr's Biological Species Concept (BSC). It classifies organisms into species. Its sorting rules are sort sexual organisms that interbreed into the same species, sort sexual organisms that do not interbreed into different species, and do not sort asexual organisms into any species. The BSC's motivating principle is to classify organisms into groups that are distinct evolutionary units, that is, groups of organisms that evolve in tandem. According to Mayr (1996, 262, 264), species are the principal units of evolution because their reproductive isolation prevents the production of incompatible gene combinations and allows adaptations to become fixed within a species. The GFA suggests that we evaluate the success of the BSC by how well sorting by interbreeding picks out groups of organisms that are distinct evolutionary groups. In other words, we evaluate whether the BSC offers natural kind classifications by how well its sorting principles achieve the classificatory program's motivating principle. This is in part a practical matter—do the sorting principles actually enable us to pick out groups of organisms in the first place?—and in part a theoretical one—do the groups that are picked out constitute distinct evolving entities? As it turns out, in many cases the sorting rules of the BSC successfully pick out distinct evolutionary units (Coyne and Orr 2004), so according to the GFA the BSC does well in classifying organisms into natural kinds.

Contrast the BSC with another species concept, the PPSC of microbiology we saw earlier. The PPSC aims to highlight stable and readily identifiable groups of microorganisms. Its motivating principle is to obtain stable microbial groups for use in microbial and medical research. Its sorting principles use various genetic parameters for sorting microbes into stable groups, such as similarities in 16S rRNA genes and DNA-DNA hybridization. According to the GFA, whether the PPSC offers natural kind classifications turns on how well its sorting principles satisfy that classificatory program's motiving principle, namely to pick out stable microbial species. According to numerous microbiologists (for example, Rosselló-Mora and Amann 2001; Stackebrandt 2006), the PPSC does achieve its aim. Thus, it too scores well on the GFA.

The BSC and the PPSC are positive cases where normative naturalism judges classificatory programs favorably. What about negative cases, where the GFA judges a classificatory program unfavorably? Consider the Phenetic Species Concept (Sneath and Sokal 1973). Its aim is to produce classifications of organisms that are free of theoretical assumptions. It sorts organisms according to overall similarity. Pheneticists construct multidimensional graphs where each dimension represents a trait and points on a graph represent sample organisms: the densest clusters of points represent species, clusters of species that are closer together on the graph represent genera, and so on. Though phenetics was popular among some biological taxonomists in the 1960s, it has fallen out of favor. The GFA properly reconstructs why it has fallen out of favor: because phenetics cannot produce classifications that achieve its overarching aim. Organisms have an indefinite number of similarities, so some similarities must be selected while most are ignored for constructing a classification. Because theoretical considerations must come into play when choosing which traits to use for constructing classifications (Hull 1970), the Phenetic Species Concept is a classificatory program whose sorting principles result in classifications that violate the school's aim of providing theory-free classifications. It thus scores poorly on the GFA.

Notice two things about these examples. First, each of the three classificatory programs discussed previously has its own overall aim, and each, according to the GFA, should be evaluated according to how well it achieves that particular aim. Second, the GFA treats natural kinds in a strikingly different way than monistic accounts of natural kinds. Those accounts set one overarching epistemic aim for evaluating *all* natural kind classifications, such as the possibility of making inductive generalizations or highlighting of the world's causal structure. The GFA is different, as the aims of classificatory programs are found in the programs themselves and can vary from program to program. Whether a program offers natural kind classifications depends on how well those classifications achieve the program's specific aims. Consequently, the GFA is sufficiently sensitive to the various aims scientists have for positing natural kind classifications, while at the same time retaining a reasonable normative component.

One might wonder why we place a functional constraint on natural kind classifications—that a natural kind classification should satisfy the aims for which it was posited. The underlying motivation is that natural kind classifications are tools for scientists to achieve various ends. We've highlighted epistemic aims, such as a classification highlighting evolutionary units or the desire to obtain stable classifications. A successful natural kind classification, we submit, should achieve the epistemic (and nonepistemic) aims it is posited to achieve. Otherwise, a natural kind classification will not serve well as the tool it was intended to be.

Before moving on, we would like to say something more general about the reasons that scientists have for producing classifications. So far, we have focused on the *epistemic* reasons scientists have for positing classifications:

for example, researchers hold that classifications should be useful for making inductive inferences or be theory free or provide stable groupings for research and so on. Although we have focused on epistemic reasons for positing classifications, we believe that scientists also typically have *nonepistemic* reasons for positing classifications (for a more detailed discussion, see Reydon and Ereshefsky 2022). Scientists routinely use contextual values, such as moral and social values, for constructing classifications (Anderson 1995; Ludwig 2014; Conix 2019). In addition, scientists use what Slater (2017) calls "cognitive values" to construct classifications. An example of such a value is the rule of avoiding "lonely categories"—categories that have only one member (Slater 2017). We won't further discuss the use of nonepistemic values in producing natural kind classifications here. However, we will suggest that cognitive and contextual values can easily be incorporated within the GFA framework, if one wanted to do so (Reydon and Ereshefsky 2022). Just as classifications should promote the epistemic aims they are posited for, one can incorporate the idea that classifications should promote the cognitive and contextual aims they are posited to achieve. We see the GFA's ability to incorporate contextual and cognitive aims as a virtue of the GFA.

4. WHAT'S NATURAL ABOUT NATURAL KINDS?

One might worry that the suggested account of natural kinds is too permissive—that merely requiring that a natural kind classification satisfy the motivating principles of a classificatory program might allow too many classifications to be natural kind classifications. The worry is that the GFA merely requires a sort of internal consistency between the aims of a classificatory program and the classifications it provides. In the preceding discussion, we restricted the set of relevant aims for classificatory programs to epistemic aims. But even under this restriction one might worry that the GFA could incorrectly designate some nonnatural kinds as natural kinds. Consider the example of "Canadian permanent resident." A political scientist might be interested in the different kinds of residents one finds in Canada, such as permanent resident, citizen resident, and various sorts of temporary residents. The aim of such a classification is to accurately describe the different kinds of residents found in Canada and study their social and political roles. The classification that refers to the category "Canadian permanent resident" satisfies that aim and thus satisfies the GFA. Nevertheless, one might argue that "Canadian permanent resident" is not a natural kind but a socially constructed kind. After all, the membership conditions for that kind were legislated by the Canadian government. Citing such an example, one might hold that the GFA provides an insufficient standard for determining if a kind is a natural kind.

To rectify this lacuna, one might turn to a standard way that philosophers distinguish natural kinds from nonnatural kinds: by adding the requirement that natural kinds exist independently of human thought or action or represent the mind-independent structure of the world (for example, Bird and Tobin 2017; Lowe 2014; Devitt 2005; Psillos 2002; and Searle 1995). Bird and Tobin (2017) provide the following version of the mind-independence requirement: "to say a kind is *natural* is to say that it corresponds to a grouping that reflects the structure of the natural world rather than the interests and actions of human beings." Despite the widespread acceptance of the mind-independence requirement, we find it too blunt of an instrument for distinguishing natural kinds from nonnatural kinds (also see Khalidi 2013; 2016; Ereshefsky 2018). To illustrate our point, we employ Kukla's (2000, ch. 3) threefold distinction among the different ways entities or categories can depend on us: material, causal, and constitutive dependence.

Consider first *material dependence*: when we make entities in the lab or the field, such as new plant species, artificially bred animals, or new chemical compounds, the members of these species, varieties, and chemical kinds, as well as the kinds themselves, come into existence due to human actions. But clearly such kinds also depend on nature as we cannot make just any organism or compound we can think of—nature constrains what is possible. The second way classificatory categories can depend on us is *causal dependence*: when kinds of people in part depend on what we think about them, those kinds can be said to depend causally on our views and actions. The "looping kinds" highlighted by Hacking (1995; 2007a) constitute prominent examples of this sort of dependence. For example, Hacking (1995) suggests that the kind "dissociative identity disorder" is affected by what medical professionals think about people diagnosed with that disorder. Depending on the state of research on the disorder, accepted diagnostic criteria, and available therapies, the kind's boundaries may shift considerably. Still, there are biochemical processes and brain states underlying the kind, such that the kind does not entirely depend on our thoughts about the kind. The third way classificatory categories can depend on us is *constitutive dependence*: when membership in a kind *entirely* depends on our thoughts and actions, we may say that the kind depends constitutively on us. Social conventionalists (for example, Woolgar 1988) and those that hold infallibilist views of social kinds (for example, Searle 1995; Thomasson 2003; Taylor 1971) discuss such constitutive kinds.

We want to suggest that there is a significant difference between kinds that materially or causally depend on us *versus* kinds that constitutively depend on us. Kinds that materially depend on us, such as genetically modified organisms and synthetic chemicals, depend on us for their initial existence. But once we create them, they take on a life of their own that we can study. We can form hypotheses about their behavior, and through empirical investigation, we can determine whether those hypotheses are correct or incorrect. Similarly, kinds that causally depend on us are affected by our thoughts but nevertheless can be empirically investigated. Here we have in mind many of the kinds studied in the social and human sciences, such as psychological kinds, sociological kinds, and economic kinds. Those kinds are affected by our psychological states and behaviors, yet we can form hypotheses about them, and empirical testing can show that those hypotheses are wrong. For instance, even though professional and societal beliefs affect the behaviors of those with dissociative personality disorder, we can form hypotheses about those behaviors and be wrong about them. On the other hand, constitutive (or conventional) kinds, such as the kind "mermaid" or the kind "Canadian permanent resident," are not open to revision on *empirical* grounds. We (in this case, users of English) implicitly define what mermaids are, and our governments legislate what permanent residents are such that those kinds' membership conditions are not based on any empirical investigations. We don't form hypotheses about the defining characteristics of such conventional kinds and subject those hypotheses to empirical testing. What mermaids or permanent residents are depends entirely on how we define those categories.

Stepping back from these examples, the significant difference between kinds that materially or causally depend on us *versus* kinds that constitutively depend on us is that kinds in the latter group depend *entirely* on human thoughts and actions, while kinds in the former two groups depend both on the world *and* on human thoughts and actions. We suggest that this *partial dependence on the world* is the factor that makes the former two groups natural (see also MacLeod and Reydon 2013; Reydon 2016), and accordingly we take this dichotomy as determining the distinction between natural and nonnatural kinds. Simply defining natural kinds as those groupings that are independent of our thoughts and actions is not an adequate way to distinguish natural from nonnatural kinds because requiring that natural kinds be independent of human thought or action leaves out important kinds in the social and human sciences as well as many areas of the natural sciences. Yet many of those disciplines provide us with an understanding of the world and the means for predicting and manipulating aspects of the world. The kinds that feature prominently in those disciplines—kinds that materially or causally depend on us—should not be ruled out from being natural kinds on the basis of their partially depending on us.

How then do we distinguish kinds that materially or causally depend on us from those that constitutively depend on us? We can do this by amending the mind-independence requirement that philosophers place on natural kinds. We suggest taking Bird and Tobin's (2017) version of that requirement and changing it to the following.

> To say a kind is *natural* is to say that it corresponds to a grouping that depends on an aspect of the world rather than *merely* on the interests and actions of human beings.

Call this the "grounding condition" on natural kinds. This is why we call our account the "Grounded Functionality Account" of natural kinds: natural kind classifications should satisfy the epistemic as well as nonepistemic aims they are posited for, and those classifications should be grounded in the world. The grounding condition is different than Bird and Tobin's mind-independence condition in a couple of ways.

First, they write that "a kind is *natural* is to say that it corresponds to a grouping that reflects the structure of the natural world" (Bird and Tobin 2017). We have dropped the word "natural" from the phrase "natural world." This is done to avoid an a priori constraining of what can be a natural kind by focusing exclusively on the nonhuman world. Because those aspects of the world that our natural kind classifications may correspond to can be human-made or not human-made, we don't want the word "natural" to rule out the former. Kinds of technical artifacts, for example, are not fundamentally different from new species of organisms that have been created by genetic technologies or by conventional breeding, or from synthetically created chemical elements (Reydon 2014). Kinds of artifacts typically are materially and causally dependent on us, but they are not *entirely* dependent on us.[2] Artifacts are not merely social conventions, and artifact kinds can be studied in the same way as kinds of natural entities can be studied: once a new kind of artifact has been designed and the first prototypes have been made, we can formulate hypotheses about them and study their behavior in practice. Much the same holds for many kinds that feature in the social sciences, such as kinds in economics. For example, we study the behavior of economic systems and the various kinds of entities featuring in them (such as consumers, money, credit institutions, and so on), even though their existence in part depends on us. Once they have been brought into existence, they take on a life of their own. We can form hypotheses about them and we can be wrong about those hypotheses. Accordingly, instead of saying that the interests and actions of humans are completely irrelevant for scientific classification, which would be a problem for the social, medical, biological, and chemical sciences, the grounding condition explicitly allows that kinds can in part but not completely depend on us.

Another way that the grounding condition differs from Bird and Tobin's (2017) criterion is that they talk in terms of "the structure of the natural world." To avoid making potentially problematic metaphysical commitments, instead of talking about *the* structure of the natural world, the grounding condition talks about aspects of the world. Doing so avoids any commitment to the world having a fundamental structure. We would like to remain agnostic about whether there is such a structure. Focusing on aspects of the world also allows us to see more clearly that any metaphysical commitment the GFA has is one of *local* metaphysics. Which aspects of the world provide the grounding of a natural kind classification depends on the aims that scientists using a classificatory program are pursuing. Consider some of the examples mentioned earlier. The sorting principles of the Biological Species Concept (BSC) turn on the assumption that interbreeding causes evolutionary units, so it is the relation between the occurrence of interbreeding among a group of organisms and that group being an evolutionary unit that needs to be grounded in the world. The term "grounding" here is used to mean the straightforward point that for a natural kind classification to be useful, its functionality (and hence the kind itself) should in some way be anchored to, based on, or supported by aspects of the world. Returning to our example, the BSC provides useful classifications of biological phenomena because it is based on a relation found in the world: that interbreeding causes the existence of evolutionary units. The BSC is a useful classificatory approach for biologists because it has latched on an aspect of the world. Similarly, the Phylo-Phenetic Species Concept (PPSC) assumes that certain genetic markers allow us to identify stable taxonomic groups.

For that approach to species to be successful, the relation it asserts—that certain genetic markers pick out stable taxonomic groups—needs to be grounded in the world. Here the grounding we are talking about is simply that the world actually contains such genetic markers as specified by the PPSC that microbiologists can use to identify stable taxonomic groups.

The notion of grounding used in this chapter should be contrasted with the notion of metaphysical grounding found in contemporary analytic metaphysics (Correia and Schnieder 2012). In metaphysical grounding, something grounds the existence of something else. For example, facts about physical particles are thought to ground facts about larger objects. Such metaphysical ground is not what we have in mind. By "grounding" in our grounding condition, we just mean that natural kind classifications make certain assumptions about the world (e.g., interbreeding causes evolutionary units) and a classification is a natural kind classification only if those assumptions are correct about the world.

Loosely put, the grounding condition says that the functionality of natural kind classifications should in part depend on the world and not merely our conceptions of it—that is, that a kind successfully serves the propose(s) for which it was posited because of the way in which it depends on the world. Note that if the way in which a kind serves its purpose depends on (or is anchored to or supported by) aspects of the world, this holds for the kind itself too. While the grounding condition requires that the way in which a kind serves its purpose depends in part on the world, this entails the requirement that the kind depends in part on the world too. Note furthermore that the grounding condition does not assume any specific way of depending on the world—it remains agnostic about the various ways in which kinds and their functionality may metaphysically depend on the world.[3] The grounding condition allows that kinds that depend materially and causally on us can be natural kinds (such as newly bred plant species or social kinds) but rules out kinds that constitutively depend on us. The grounding condition makes sense in the abstract. Natural kind classifications are tools for gaining knowledge about the world—such classifications are made by us in the context of classificatory programs that have specific epistemic (or other) aims. To serve as such tools, natural kind classifications should depend on the world and not merely on our conceptions of it. Otherwise, they will not allow us to successfully investigate and manipulate the world.[4]

With the grounding condition articulated, let us mention how the two parts of our account of natural kinds—the grounding condition and the functionality condition—fit together. The functionality condition, as discussed in the previous section, says that natural kind classifications should satisfy the epistemic aims (or other sorts of aims) they are posited for. The grounding condition asserts that a natural kind classification should be grounded in the world. What determines how a natural kind classification should be grounded is the epistemic (or other) aim for which the classification was posited. That is, the intended function of the classification sets out which aspects of the world should ground a natural kind classification. Returning to our well-worn species concept examples, the BSC aims to give classifications of evolutionary units. The BSC asserts that interbreeding is a factor that underlies evolutionary units. Therefore, the relation that needs to be grounded in the world for BSC classifications to be natural kind classifications is that interbreeding does indeed cause the existence of evolutionary units. Turning to our example from microbiology, proponents of the PPSC assert that certain genetic markers identify stable taxonomic units. Consequently, the relation that needs to be grounded in the world for the PPSC to provide natural kind classifications is that the highlighted genetic markers do indeed pick out stable groups of organisms.[5] In both classificatory examples, it is the function of a classification—the aim for which it is posited—that determines how a natural kind classification should be grounded in the world.

Let's take stock of where we are in the search for a more practice-oriented account of natural kinds. In the second section of this chapter, we saw that philosophical accounts of natural kinds fail to properly interact with and account for successful classificatory practices in science. They are either based on a priori reasoning and thus irrelevant to actual classificatory practices, or they overlay a single epistemic aim on why all scientists posit natural kind classifications, when in fact scientists posit natural kind classifications for a variety of reasons. In an attempt to offer a more practice-oriented account of natural kinds we suggested an approach to natural kinds inspired by Laudan's normative naturalism and Woodward's functional account of causal reasoning: we should judge a natural kind classification according to how well it satisfies the epistemic (or other) reasons it was posited for. But then there was the worry that this is a too permissive of an approach to natural kinds. So we suggested that natural kind classifications should satisfy the grounding condition: natural kind classifications and the functionality of the kinds it recognizes should in part depend on the world and not *only* on our conceptions of it.

Let's put the grounding condition and the functionality condition of the GFA together. A natural kind classification should satisfy the functionality condition of the GFA—that is, a natural kind classification should achieve the epistemic aims (or other aims) it was posited for. Furthermore, the way that a natural kind classification satisfies that function should be grounded in the world. We take the grounding condition and the functionality condition each to be necessary conditions for classifications to be natural kind classifications. To meet the grounding condition, we need to know which aspects of the world a kind or classification is supposed to highlight—which is what the functionality condition of our account tells us. In other words, without a classificatory program that specifies the basis on which entities are to be grouped together into kinds, there is no way to examine whether and, if so, how the kinds are grounded in the world. This is why both the grounding condition and the functionality condition are necessary conditions. Together they constitute a jointly sufficient condition for determining which classifications are natural kind classifications.

The grounding condition is an all or nothing condition: we ask if a natural kind classification is appropriately grounded in the world. That is, we ask whether a natural kind classification at least in part appropriately depends on the world and not merely on our conceptions of it. When it comes to the functionality condition, it is reasonable to think there is a sliding scale. How well a classification may achieve its aim may come in degrees. For instance, the stability of PPSC classifications, the aim of such classifications, might not be an all or nothing affair but may come in degrees. Putting this all together, the GFA asserts that a natural kind classification must be grounded in the world and must satisfy the epistemic (or other) aims it was posited for, though how well it satisfies those aims may come in degrees. The GFA, we submit, is naturalistic enough to capture actual classificatory practices in science, and at the same time it has a significant normative component.

5. RECENT SKEPTICISM ABOUT NATURAL KINDS

Recently, some philosophers have voiced skepticism concerning philosophical research on natural kinds (Hacking 2007b; Ludwig 2018). We believe that the GFA can go some way in answering such skepticism. Consider what Hacking has to say about philosophical research on natural kinds. According to Hacking, modern philosophical research on natural kinds began as a "rosy dawn" with the work of Mill and Whewell in the nineteenth century.

But in the late twentieth and early twenty-first centuries, that work entered a "scholastic twilight" (2007b, 203). Hacking tells us that philosophical research on natural kinds is now "a slew of distinct analyses directed at unrelated projects" (2007b, 203). Moreover, Hacking argues that philosophical research on natural kinds focuses on "an inbred set of degenerating problems that have increasingly little to do with issues that arise in a larger context," where "a larger context" refers to classificatory projects in science and elsewhere (2007b, 229).

We believe that a fruitful way to answer Hacking's pessimism is to refocus philosophical work on the topic by moving away from a priori considerations regarding natural kinds and more carefully studying classificatory practices in science. We suggest that practice-oriented philosophical analyses of natural kinds *are* related and that those analyses *do* address issues that arise in a larger context (MacLeod and Reydon 2013).[6] In particular, we believe that the grounding condition captures a common concern among practice-oriented philosophers who work on natural kinds as well as among those scientists who worry about what makes a classification natural. Those philosophers and scientists attempt to articulate how natural kind classifications should (at least in part) depend on the world and not entirely on our conceptions of it. They just disagree on the ways that natural kind classifications should be grounded in the world.

Consider two opposing practice-oriented philosophical approaches to natural kinds, Boyd's (1999a; 1999b; 2003) homeostatic property cluster theory and Slater's (2013; 2015) stable property cluster theory. For Boyd, natural kinds have two components. They are groups of entities that have co-occurring clusters of properties that sustain successful induction. Furthermore, that co-occurrence of properties is underwritten by causal mechanisms. For instance, *Canis familiaris* is a natural kind on Boyd's account because dogs have a number of co-occurring properties, such as having four legs and having a tail, and the occurrence of those properties is caused by such homeostatic mechanisms as genealogy and shared developmental pathways. Boyd requires that kinds achieve "the accommodation of inferential practices to relevant causal structures" (1999a, 159). Boyd's account is a realist one in the sense that it requires that causal structures sustain natural kinds. Slater, on the other hand, does not require that natural kinds be sustained by causal structures or any particular mechanism. Like Boyd, Slater requires that natural kinds are associated with stable clusters of properties that can be used for induction. But for Slater bare stability is all that is required for natural kinds: natural kinds are simply stable clusters of proper-ties that underwrite induction, no matter how that stability is realized.

Despite their differences, both Boyd and Slater agree that natural kinds should in some way be grounded in the world and should not merely be the result of our conceptions. They just disagree on *how* natural kinds should be grounded. For Boyd, natural kinds are grounded in clusters of stable prop-erties and causal mechanisms. For Slater, the grounding of natural kinds just depends on there being stable property clusters in the world. Boyd's and Slater's accounts are not, as Hacking puts it, "unrelated projects": they both want to ground natural kinds in the world; they just disagree on which fea-tures of the world provide that grounding.

The assumption that natural kind classifications should be grounded in the world is also found in Hacking's own work on human kinds. Hacking (1991; 1995) initially drew a division between natural and human kinds. Hu-man kinds, such as dissociative identity disorder, depend in part on our conceptions of those kinds, whereas natural kinds, such as silver, do not de-pend on our conceptions. Hacking (2002) revised his view such that the salient division is between "indifferent kinds" and "interactive kinds" (or "looping kinds"). Indifferent kinds (for example, silver) are unaffected by what we think about them, whereas interactive kinds (for example, disso-ciative identity disorder) are affected by what we think about them. Although the word "natural" has fallen out, both kinds of kinds depend in part on the world and not entirely on our conceptions. Furthermore, Hacking (1999, 126–27) clearly distinguishes indifferent and interactive kinds from con-stituent or conventional kinds. He offers "satanic ritual abuse" as an exam-ple of a constituent kind that is not grounded in the world but is merely found in our conceptions. So even in this account of human kinds by a vo-cal critic of the concept of natural kinds, we find the distinction between kinds that in part depend on the world versus kinds that entirely depend on our conceptions of the world. In other words, we see the grounding condition at work even in Hacking's writings.[7] Contrary to Hacking's claim that phil-osophical accounts of natural kinds are "unrelated projects," we see that several different accounts of natural kinds (Boyd's, Slater's, and Hacking's own accounts) hold that natural kinds are grounded in the world and not merely in our conceptions of the world.

Let's turn to Hacking's charge that philosophical analyses of natural kinds are "an inbred set of degenerating problems that have increasingly lit-tle to do with issues that arise in a larger context" (2007b, 229). For a larger context, let's turn to how biologists characterize the difference between natural and nonnatural classifications. Consider the works of several biological taxonomists: Mayr (1982), Panchen (1992), and Baum and Smith (2013). These biologists characterize the history of biological taxonomy as a search for criteria that distinguish natural from nonnatural classifications and recount that history in terms of how criteria for natural classifications vary over time. These biologists aim to promote their favored school of taxonomy, so the history told is one where previous taxonomic schools allegedly focused on the wrong criteria for natural classifications. Mayr is a promoter of the taxonomic school of evolutionary taxonomy, while the other authors subscribe to cladism. Mayr holds that natural classifications should capture both propinquity of descent and adaptive variation, in other words, classifications should sort organisms into taxa according to their phylogeny *and* their adaptive differences. Cladists, on the other hand, argue that *only* propinquity of descent is the aspect of the world that should be captured in natural classifications. Cladists criticize evolutionary taxonomy for relying on what they see as subjective measures of adaptive difference. Stepping back from these details, cladists and evolutionary taxonomists agree that an overall aim of their discipline is to distinguish natural from nonnatural classifications. They agree that natural classifications should be grounded in the world and not our mere conceptions of it. They just disagree on which aspects of the world ground natural classifications in biology.

The same can be said of two classificatory programs that we looked at earlier. Supporters of the BSC and supporters of the PPSC agree that natural classifications should be grounded in the world: supporters of the BSC focus on interbreeding causing evolutionary units, whereas supporters of the PPSC focus on certain genetic markers picking out stable taxonomic units. Supporters of the BSC and supporters of the PPSC thus agree that natural kind classifications should be grounded in the world, but disagree on which aspects of the world are the relevant ones for grounding natural classifications.

This brief survey of philosophers working on natural kinds and biologists interested in what makes a classification natural undermines Hacking's charges against natural kind research. Practice-oriented philosophers working on theories of natural kinds are interested in how natural kinds are grounded in the world and not merely in our conceptions of it. Similarly, biological taxonomists interested in what makes classifications natural ones disagree over *how* their classifications should be grounded in the world, but they nevertheless agree *that* natural classifications should be grounded in the world. The project of investigating natural kinds, we submit, is more unified than Hacking claims (see also MacLeod and Reydon 2013, 91).

Before concluding, let us briefly address a more recent article that holds a view similar to Hacking's. David Ludwig (2018) argues that standard philosophical accounts of natural kinds tend to focus on particular and limited reasons scientists have for producing classifications. According to Ludwig, standard philosophical accounts of natural kinds "privilege some dimension of nonarbitrariness over others and can therefore lead to an unnecessarily narrow analysis of classificatory practices" (2018, 47). We agree with Ludwig that standard philosophical accounts neglect the actual variety of reasons that scientists have for classifying entities under investigation. However, we disagree with the conclusion he derives from this. Ludwig believes that we should let go of the concept of natural kind and instead just focus on the different ways that scientists offer nonarbitrary classifications. We disagree with this conclusion because we believe that the grounding condition captures what various philosophers and scientists aspire to when they talk about *natural* classifications. In addition, the grounding condition does not face the problem that Ludwig attributes to other accounts of natural kinds: it does not unnecessarily limit analyses of classificatory practices. Furthermore, the grounding condition does something that Ludwig thinks an account of natural kinds should do, namely, have a normative aspect, which rule outs "wildly pathological" classifications (such as the group of all animals born on a Tuesday) and "scientifically defunct" classifications (such as hysteria) (Ludwig 2018, 47; see also Franklin-Hall 2015, 926).

6. THE GFA'S PRINCIPAL VIRTUES

Let us conclude by highlighting that the Grounded Functionality Account of natural kinds has several virtues that other accounts lack. We argued that a new account of natural kinds is required because available accounts fail on one or more of the following counts. First, many accounts are based on a priori assumptions about the nature of natural kinds that cause those accounts to either neglect or be irrelevant to important aspects of scientific practice. Second, many accounts acknowledge just one epistemic aim for which all natural kind classifications are posited; consequently, they miss the diversity of classificatory practices found in science. Third, most accounts fail to acknowledge that legitimate nonepistemic aims may be important in the positing of scientific classifications. Here, again, aspects of actual classificatory practice in science are overlooked.

Our alternative, the GFA, is naturalistic enough to be relevant to actual classificatory practices in science and avoid the three counts of failure mentioned previously. The GFA is not an a priori approach to natural kinds, nor does it focus on only one epistemic aim, nor does it exclude nonepistemic aims. Thus it is naturalistic enough to capture the various classificatory practices found in the sciences, unlike standard philosophical theories of natural kinds. Furthermore, the GFA is not only sufficiently naturalistic to capture the diversity of classificatory practices found in science, it also has a significant normative component. Through the application of the functionality condition, what counts as a natural kind classification is constrained by satisfying the epistemic (or other) goals for which a classification is posited. In addition, what is a natural kind classification is further constrained by the grounding condition. Finally, the GFA entails that Hacking's and Ludwig's pessimism about philosophical work on natural kinds is too hasty. Practice-oriented philosophical theories of natural kinds are not an array of unrelated projects that have no connection to scientific classification. There is a common overarching aim of philosophical work on natural kinds and scientific work on natural classifications, namely that natural classifications should be grounded in the world and not merely our conceptions of it. Like Hacking and Ludwig, we are pessimistic about many of the available philosophical theories of natural kinds, but unlike those authors we are optimistic about the usefulness of 'natural kind' as a philosophical concept.

NOTES

1. The view that classifications are posited to serve particular aims is not new. It was prominently argued for by, among others, Dupré (1993). What is new in our account is the normative aspect—success in achieving an aim for which a classification is posited is part of what makes a classification a natural kind classification.

2. Because of this, Reydon (2014) argues that artifact kinds are not natural kinds in any *traditional* sense of the term, but not artificial (i.e., conventional) kinds either. The GFA recognizes artifact kinds as natural kinds.

3. In particular, the GFA does not require such dependence to always be causal dependence. Even though, as explained previously, the GFA does not involve metaphysical grounding in the sense of current analytic metaphysics,

the GFA still aligns with parts of the current literature on metaphysical grounding (see, for example, Bliss and Trogdon 2016) in not presupposing that there is only one way of depending on the world or requiring that all dependence is causal but remains agnostic about the question in what ways kinds and their functionality may depend on the world. By allowing this, the GFA avoids becoming bogged down by deeper metaphysical issues.

4. The grounding condition for natural kinds is far from new. Locke, for instance, distinguishes between two factors that contribute to natural kind classifications: the "workmanship of nature" and the "workmanship of the understanding" (or the "workmanship of men") (Reydon 2016, 62). For Locke, both are important aspects of classifications: while it is nature that makes things similar and different to various degrees, it is we who use these similarities and differences to group things into kinds that can be used for various purposes (*Essay,* III.III.§13, III.VI.§37). According to Locke, purely nominal kinds, that is, those kinds that depend only on how we define kind terms, can be used in practice for communicative purposes. In scientific investigations, by contrast, we look for kinds that to some extent depend on the "workmanship of nature." In Lockean terms, we would say that a kind fails the grounding condition and is not natural if it depends *only* upon the "workmanship of men" and not in any way on the "workmanship of nature."

5. Obtaining such stability is not a trivial manner. That is why the PPSC uses three different types of genetic markers to identify species. Stability is achieved by triangulating these three markers.

6. By practice-oriented philosophical theories of natural kinds, we mean those theories that aim to be consistent with and learn from actual classificatory practices in science.

7. Note that Hacking did not explicitly conceive of his work as involving such a thing as the grounding condition formulated here.

REFERENCES

Anderson, E. 1995. "Knowledge, Human Interests, and Objectivity in Feminist Epistemology." *Philosophical Topics* 23: 27–58.

Baum, D., and S. Smith. 2013. *Tree Thinking: An Introduction to Phylogenetic Biology.* Greenwood Village, Colo.: Roberts and Company Publishers.

Bird, A., and E. Tobin. 2017. "Natural Kinds." In *The Stanford Encyclopedia of Philosophy* (spring 2017 edition), edited by E. Zalta. http://plato.stanford.edu /archives/spr2017/entries/natural-kinds/.

Bliss, R., and K. Trogdon. 2016. "Metaphysical Grounding." In *The Stanford Encyclopedia of Philosophy* (winter 2016 edition), edited by E. Zalta. https://plato.stanford.edu/archives/win2016/entries/grounding/.

Boyd, R. 1999a. "Homeostasis, Species, and Higher Taxa." In *Species: New Interdisciplinary Essays,* edited by Robert A. Wilson, 141–85. Cambridge, Mass.: MIT Press.

Boyd, R. 1999b. "Kinds, Complexity and Multiple Realization." *Philosophical Studies* 95: 67–98.

Boyd, R. 2003. "Finite Beings, Finite Goods: The Semantics, Metaphysics and Ethics of Naturalist Consequentialism." *Philosophy and Phenomenological Research* 66: 505–53.

Bryant, A. 2017. "Keep the Chickens Cooped: The Epistemic Inadequacy of Free Range Metaphysics." *Synthese* (April): 1–21.

Callender, C. 2011. "Philosophy of Science and Metaphysics." In *The Continuum Companion to the Philosophy of Science,* edited by S. French and J. Saatsi, 33–54. London: Continuum.

Conix, S. 2019. "Radical Pluralism, Classificatory Norms and the Legitimacy of Species Classifications." *Studies in History and Philosophy of Biological and Biomedical Science* 73: 27–34.

Correia, F., and B. Schnieder, eds. 2012. *Metaphysical Grounding: Understanding the Structure of Reality.* Cambridge: Cambridge University Press.

Coyne, J. A., and H. A. Orr. 2004. *Speciation.* Sunderland, Mass.: Sinauer.

Craver, C. F. 2009. "Mechanisms and Natural Kinds." *Philosophical Psychology* 22: 575–94.

Devitt, M. 2005. "Scientific Realism." In *The Oxford Handbook of Contemporary Philosophy,* edited by F. Jackson and M. Smith, 766–90. Oxford: Oxford University Press.

Dupré, J. A. 1993. *The Disorder of Things: Metaphysical Foundations of the Disunity of Science.* Cambridge, Mass.: Harvard University Press.

Ellis, B. 2001. *Scientific Essentialism.* Cambridge: Cambridge University Press.

Ereshefsky, M. 2001. *The Poverty of the Linnaean Hierarchy: A Philosophical Study of Biological Taxonomy.* Cambridge: Cambridge University Press.

Ereshefsky, M. 2018. "Natural Kinds, Mind-Independence, and Defeasibility." *Philosophy of Science* 85: 845–56.

Ereshefsky, M., and T. A. C. Reydon. 2015. "Scientific Kinds." *Philosophical Studies* 172: 969–86.

Franklin-Hall, L. 2015. "Natural Kinds as Categorical Bottlenecks." *Philosophical Studies* 172: 925–48.

Hacking, I. 1991. "A Tradition of Natural Kinds." *Philosophical Studies* 61: 109–26.

Hacking, I. 1995. "The Looping Effects of Human Kinds." In *Causal Cognition: A Multidisciplinary Debate,* edited by D. Sperber, D. Premack, and A. J. Premack, 351–83. Oxford: Clarendon Press.

Hacking, I. 1999. *The Social Construction of What?* Cambridge, Mass.: Harvard University Press.

Hacking, I. 2002. "Inaugural Lecture: Chair of Philosophy and History of Scientific Concepts at the Collège de France, 16 January 2001." *Economics and Society* 31: 1–14.

Hacking, I. 2007a. "Kinds of People: Moving Targets." *Proceedings of the British Academy* 151: 285–318.

Hacking, I. 2007b. "Natural Kinds: Rosy Dawn, Scholastic Twilight." In *Philosophy of Science (Philosophy—Royal Institute of Philosophy Supplement 61),* edited by A. O'Hear, 203–39. Cambridge: Cambridge University Press.

Hawley, K., and A. Bird. 2011. "What Are Natural Kinds?" *Philosophical Perspectives* 25: 205–21.

Hendry, R. 2010. "The Elements of Conceptual Change." In *The Semantics and Metaphysics of Natural Kinds,* edited by H. Beebee and N. Sabbarton-Leary, 137–58. London: Routledge.

Hull, D. L. 1970. "Contemporary Systematic Philosophies." *Annual Review of Ecology and Systematics* 1: 19–54.

Khalidi, M. A. 2013. *Natural Categories and Human Kinds: Classification in the Natural and Social Sciences.* Cambridge: Cambridge University Press.

Khalidi, M. A. 2016. "Mind-Dependent Kinds." *Journal of Social Ontology* 2: 223–46.

Khalidi, M. A. 2018. "Natural Kinds as Nodes in Causal Networks." *Synthese* 195: 1379–96.

Kukla, A. 2000. *Social Constructivism and the Philosophy of Science.* New York: Routledge.

Laudan, L. 1987. "Progress or Rationality? The Prospects for Normative Naturalism." *American Philosophical Quarterly* 24: 19–33.

Laudan, L. 1990. "Normative Naturalism." *Philosophy of Science* 57: 44–59.

Lowe, E. J. 2006. *The Four-Category Ontology: A Metaphysical Foundation for Natural Science.* Oxford: Oxford University Press.

Lowe, E. J. 2014. "How Real Are Artefacts and Artefact Kinds?" In *Artefact Kinds: Ontology and the Human-Made World,* edited by M. P. M. Franssen, P. Kroes, T. A. C. Reydon, and P. E. Vermaas, 17–26. Dordrecht: Springer.

Ludwig, D. 2014. "Disagreement in Scientific Ontologies." *Journal for General Philosophy of Science* 45: 119–31.

Ludwig, D. 2018. "Letting Go of 'Natural Kind': Towards a Multidimensional Framework of Non-arbitrary Classification." *Philosophy of Science* 85: 31–52.

MacLeod, M., and T. A. C. Reydon. 2013. "Natural Kinds in the Life Sciences: Scholastic Twilight or New Dawn?" *Biological Theory* 7: 89–99.

Magnus, P. D. 2012. *Scientific Enquiry and Natural Kinds: From Planets to Mallards.* Basingstoke: Palgrave Macmillan.

Mayr, E. 1982. *The Growth of Biological Thought: Diversity, Evolution, and Inheritance.* Cambridge, Mass.: Harvard University Press.

Mayr, E. 1996. "What Is a Species, and What Is Not?" *Philosophy of Science* 63: 262–77.

Panchen, A. L. 1992. *Classification, Evolution, and the Nature of Biology.* Cambridge: Cambridge University Press.

Papineau, D. 2015. "Naturalism." In *The Stanford Encyclopedia of Philosophy* (fall 2015 edition), edited by E. Zalta. https://plato.stanford.edu/archives/win 2016/entries/naturalism/.

Psillos, S. 2002. *Causation and Explanation.* Montreal: McGill-Queen's University Press.

Putnam, H. 1975. "The Meaning of 'Meaning.'" In *Language, Mind, and Knowledge (Minnesota Studies in the Philosophy of Science, Vol. VII),* edited by K. Gunderson, 131–93. Minneapolis: University of Minnesota Press.

Reydon, T. A. C. 2006. "Generalizations and Kinds in Natural Science: The Case of Species." *Studies in History and Philosophy of Biological and Biomedical Sciences* 37: 230–55.

Reydon, T. A. C. 2010a. "How Special Are the Life Sciences? A View from the Natural Kinds Debate." In *The Present Situation in the Philosophy of Science,* edited by F. Stadler, 173–88. Dordrecht: Springer.

Reydon, T. A. C. 2010b. "Natural Kind Theory as a Tool for Philosophers of Science." In *EPSA—Epistemology and Methodology of Science: Launch of the European Philosophy of Science Association,* edited M. Suárez, M. Dorato, and M. Rédei, 245–54. Dordrecht: Springer.

Reydon, T. A. C. 2014. "Metaphysical and Epistemological Approaches to Developing a Theory of Artifact Kinds." In *Artefact Kinds: Ontology and the Human-Made World,* edited by M. P. M. Franssen, P. Kroes, T. A. C. Reydon, and P. E. Vermaas, 125–44. Dordrecht: Springer.

Reydon, T. A. C. 2016. "From a Zooming-In Model to a Co-creation Model: Towards a More Dynamic Account of Classification and Kinds." In *Natural*

Kinds and Classification in Scientific Practice, edited by C. E. Kendig, 59–73. New York: Routledge.

Reydon, T. A. C., and M. Ereshefsky. 2022. "How to Incorporate Non-Epistemic Values into a Theory of Classification." *European Journal for Philosophy of Science* 12: 4.

Ridley, M. 1993. *Evolution.* Oxford: Blackwell Scientific Publications.

Rosselló-Mora, R., and R. Amann. 2001. "The Species Concept for Prokaryotes." *FEMS Microbiology Reviews* 25: 39–67.

Samuels, R. 2009. "Delusions as a Natural Kind." In *Psychiatry as Cognitive Neuroscience: Philosophical Perspectives,* edited by M. Broome and L. Bortolotti, 49–79. Oxford: Oxford University Press.

Searle, J. 1995. *The Construction of Social Reality.* New York: Free Press Publishing.

Slater, M. H. 2013. *Are Species Real? An Essay on the Metaphysics of Species.* Basingstoke: Palgrave Macmillan.

Slater, M. H. 2015. "Natural Kindness." *British Journal for the Philosophy of Science* 66: 375–411.

Slater, M. H. 2017. "Pluto and the Platypus: An Odd Ball and an Odd Duck—On Classificatory Norms." *Studies in History and Philosophy of Science* 61: 1–10.

Sneath, P. H. A., and R. R. Sokal. 1973. *Numerical Taxonomy: The Principles and Practice of Numerical Classification.* San Francisco: W. H. Freeman and Company.

Stackebrandt, E. 2006. "Defining Taxonomic Ranks." In *Prokaryotes: A Handbook on the Biology of Bacteria, Vol. 1,* edited by M. Dworkin, 29–57. New York: Springer.

Taylor, C. 1971. "Interpretation and the Sciences of Man." *Review of Metaphysics* 25: 3–51.

Thomasson, A. 2003. "Realism and Human Kinds." *Philosophy and Phenomenological Research* 67: 580–609.

Wilson, R. A., N. J. Barker, and I. Brigandt. 2007. "When Traditional Essentialism Fails: Biological Natural Kinds." *Philosophical Topics* 35: 189–215.

Woodward, J. 2014. "A Functional Account of Causation; or, a Defense of the Legitimacy of Causal Thinking by Reference to the Only Standard That Matters—Usefulness (as Opposed to Metaphysics or Agreement with Intuitive Judgment)." *Philosophy of Science* 81: 691–713.

Woolgar, S. 1988. *Science, the Very Idea.* London: Tavistock Publications.

CONTRIBUTORS

William C. Bausman is postdoctoral fellow as part of the interdisciplinary "Evolving Language" project by the Swiss National Science Foundation National Centres of Competence in Research. He is based in the Department of Philosophy at the University of Zurich.

Janella K. Baxter is assistant professor in the Department of Psychology and Philosophy at Sam Houston State University.

Richard Creath is President's Professor of Life Sciences and Philosophy and director of the program in history and philosophy of science at Arizona State University. He is the general editor of *The Collected Works of Rudolf Carnap.*

Marc Ereshefsky is professor of philosophy at the University of Calgary.

Marie I. Kaiser is professor for philosophy of science at the Department of Philosophy and the Institute for Interdisciplinary Studies of Science (I2SoS) at Bielefeld University.

Oliver M. Lean earned his PhD in Philosophy from the University of Bristol in 2016 and has held postdoctoral positions at the University of Calgary in Canada and Université Catholique de Louvain in Belgium. He currently works in industry as a data scientist.

Thomas A. C. Reydon is professor of philosophy of science and technology in the Institute of Philosophy and the Centre for Ethics and Law in the Life Sciences (CELLS) at Leibniz Universität Hannover, as well as associated faculty in the Socially Engaged Philosophy of Science Group (SEPOS) at

Michigan State University. He is coeditor in chief of the *Journal for General Philosophy of Science* and a fellow of the Linnean Society of London.

Lauren N. Ross is associate professor in the logic and philosophy of science department at the University of California, Irvine.

Rose Trappes is a postdoctoral research fellow at Egenis, the Centre for Study of the Life Sciences, at the University of Exeter, UK.

Marcel Weber is professor of philosophy of science in the department of philosophy, University of Geneva, Switzerland, and a member of the German National Academy of Sciences Leopoldina. His publications include *Philosophy of Experimental Biology* (2005) and *Philosophy of Developmental Biology* (2022).

William C. Wimsatt was Winton Professor of Liberal Arts at the University of Minnesota, now emeritus. He is author of *Re-Engineering Philosophy for Limited Beings: Piecewise Approximations to Reality* (Harvard 2007) and coeditor of *Characterizing the Robustness of Science: After the Practice Turn in Philosophy of Science* (2012) and *Developing Scaffolds in Evolution, Culture, and Cognition* (2013).

INDEX

actualism, 42, 52

adaptation, xxxi, 10, 13, 28, 34, 50, 92, 98; basis of, 37; biological, xix, xx, 41; definitions of, 56n7; ecological, 35, 45, 56n15, 57n15; evolutionary, 45, 51, 57n15; framework, 29, 41, 44, 48, 52; functional, 33, 37, 41, 44; heuristics and, 11; inference, 32, 49; proxy by, 37–38; structures, 14, 15

Adler, Mortimer J., 66–68

aetiology, 217, 218

affordances, 86, 88–89, 92, 102, 108, 110n1

aggression, 122, 137 (fig.), 138, 140

allopatric speciation, 243

alternative splicing, 160–61

Alzheimer's disease, 231n14

amino acids, 13, 160, 169, 174n4, 175n15

Andersen, Holly, 86

Angraecum sesquipedale, 55n1

Antarctic fur seal (*Arctocephalus gazella*), 123

Antennapedia genes, 197

anti-realism, realism and, 55

ants (*Cataglyphis*), navigation by, 94

architecture: biological, 3, 4, 6, 8; causal, 213, 214 (fig.), 220, 221, 223, 224, 225, 228; common genetic, 221; structural, 44

Arias, A. M., 193

Arrhenius, Svante, 63

Aspen Institute, 66

Aufbau (Carnap), 65, 71, 80n8

autism spectrum disorder, 221

Autzen, Bengt, 203n15

Avigad, Jeremy, 14

bacteria, 126, 128, 134, 196, 197, 230, 242

Baum, D., 258

Bausman, William, viii, x, xxxi

Baxter, Janella, viii, xiv, xxxiv–xxxv, 203n15

Bechtel, W., 28, 87, 106

beetles, phenotypes of, 128

behavior, 106, 131, 140; animal, 21n2, 94; brain and, 228; computer simulations of, 22n6; human, 21n2, 22n6; methodological, 20–21; psychological, 250; withdrawal, 203n13

Bennett, Karen, 79n2